H F

ANTENNA TOPICS

1-12

by K3MT

Culpeper, VA

SECOND EDITION - FEBRUARY 2017

Copyright © 2003, 2017 by Michael J. Toia

ISBN: 978-0-9600859-1-0

16206 Glenhollow Ct   Culpeper, VA 22701   k3mt@arrl.net

PREFACE

First licensed as a novice in 1952 and shortly thereafter upgraded to general class, I've been fascinated by radio antennas for better than half a century. Various mentors have taught me, at times with, and at times without, tuition, some teachings retained, and some not. The retained part has guided me through profuse experimentation with radio antennas and propagation. Of the hundreds of designs I have built and/or used over the years, I select here a handful of useful, easy, and sometimes downright unusual ones that work well for amateur radio. Some have never before been published. I pass these on, in an effort to preserve the knowledge entrusted to me by teachers so that a part of their wisdom, while filtered through my brain, may not perish.

As a professional antenna engineer, I've designed, installed, and used antennas while in the U.S. Army Signal Corps, in Voice of America's Antennas and Propagation Shop, as a Principal Member of the Technical Staff with one of Northrop Grumman's various divisions, and as a subject matter expert on the subject working for the US Government. I am now retired.

This is not a treatise on antenna *theory*. It's more a cook book with recipes for antennas that are easy to build, and that *work*. For a discussion of theory, consult other sources, possibly my companion book, *RADIO ANTENNAS*.

Author's notes

Grammar and an apology

One once used "*passive voice*" construction in technical writing. We would say, "The tower was made of aluminum." It matters not *who* made the tower, only of what it is made. Today's technical editors insist on active voice, and cast this sentence as, " *I* made the tower of aluminum." This is misleading. Perhaps Rohn, or TelRex, or Rockwell, or Harris, or a friend, made the tower. It matters only that the tower is aluminum. The beauty of passive voice is it focuses on what was done, not *who done it*.

The pressures of technical editors have been heeded. Er, I've been bending to the pressures of my editors. Though I'm urged to do so, I've found use of "I" awkward. The result seems too braggardly. Your forgiveness is begged: go needle *your* technical editor. Mine demanded, "Don't use passive voice in your writing." My answer? ***It is done***!

Scaling

Antenna theory rests on a set of linear differential equations, the famous *Maxwell's Equations*. What does that mean to you? Antennas scale linearly. A design of a yagi, quad, or any antenna at 50.1 MHz, can be used at 21.2 MHz if you multiply every dimension by the ratio 50.1 / 21.2. *All* dimensions, including diameters, must be scaled. But diameter scaling is often ignored with little, if any, noticeable effect, provided ridiculous dimensions are avoided. A dipole antenna made of thin (#12) wire for 80 meters will work as well if scaled to make a dipole of the same wire at ten meters, but will probably not work the same if scaled to 30 GHz, where the wire is no longer "thin."

Reciprocity

I often discuss antennas as transmitters, and at other times as receivers. It makes no difference in the pattern, but *does* influence the size of components. Transmit antennas must withstand the much higher currents induced by the transmitter, while receive antennas can be made of fairly flimsy materials and yet work.

National Electrical Code

Amateur antennas are covered by electrical codes. Article 810-C of NFPA70 (The National Electrical Code) has a lot to say about the matter. In general, it recommends wire antennas be made of #14 hard drawn copper wire or heavier, for spans of up to 150 feet. There are a lot of other recommendations in the code. You can buy a copy of the code at many hardware, building supply, and electrical supply stores.

Antenna Patterns

Radiation patterns shown in this book are calculated, using the Reflection Coefficient Approximation. HF antennas are often located within a wavelength or so of the ground. The pattern is as seen from a great distance, well over twenty wavelengths, and at a point many, many wavelengths above ground: i.e., the ionosphere. None shown are actual measured results, as I do not have facilities to "fly" the pattern.

To KG4ABD, who married into the hobby, later worked at the FCC's Amateur and Citizens Division, and has supported me through the years in my quest for yet another radio antenna. Who could ask for a finer XYL?

INDEX

NOTES

THE HALF-WAVE DIPOLE

EVERYBODY'S REFERENCE ANTENNA

One of the most useful simple antennas is a single wire, cut in the middle, with a feedline attached at that point. The length of this wire is a half-wavelength, calculated by the simple equation   length (feet) = 468 / f (MHz) [1] or length (inches) = 5616 / f (MHz). The feed point impedance is 76 ohms when this antenna is mounted in outer space. However, mounted closer to the earth, its impedance may be a bit less, and a good match to 50 ohm transmission line is often achieved.

The antenna is balanced: it would work best with a balanced 50 or 75 ohm transmission line. However, most RF transmission line is coaxial, so some sort of balancing mechanism ought to be used at the antenna feed. This is not a critical issue, though. Many, many dipoles fed with simple 50 ohm RG-58 or RG-8, or 75 ohm RG-59 or RG-11 coax, have been installed and their operation is hard to distinguish from dipoles using 1:1 baluns. In 1971-73, I worked DXCC using a pair of 3-band vertical dipoles slung in an oak tree, and both of the dipoles were fed with ordinary RG-59, without baluns.

The center of the wire is cut and attached to an insulator. The insulator gap should be less than about 1% of a wavelength, or not much more. Feedline - often coaxial cable - is attached, the center conductor to one half of the dipole, and the shield to the other half.

----

[1] The true half-wave dimension is 492 / f(MHz) but electrons do not move quite at the speed of light, so the antenna needs to be a bit shorter to be at resonance. This shortening is often called "end effect."

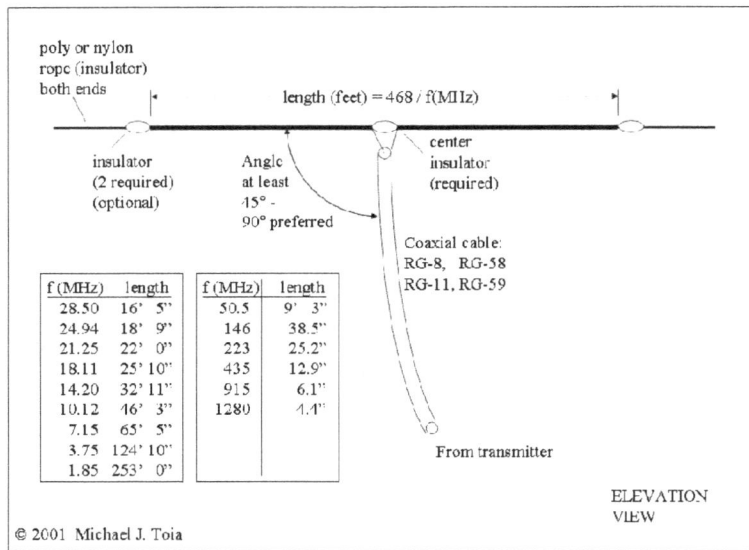

poly or nylon rope (insulator) both ends	length (feet) = 468 / f(MHz)	

insulator (2 required) (optional)

Angle at least 45° - 90° preferred

center insulator (required)

Coaxial cable: RG-8, RG-58 RG-11, RG-59

From transmitter

ELEVATION VIEW

f (MHz)	length	f (MHz)	length
28.50	16' 5"	50.5	9' 3"
24.94	18' 9"	146	38.5"
21.25	22' 0"	223	25.2"
18.11	25' 10"	435	12.9"
14.20	32' 11"	915	6.1"
10.12	46' 3"	1280	4.4"
7.15	65' 5"		
3.75	124' 10"		
1.85	253' 0"		

© 2001 Michael J. Toia

THE HALF-WAVE DIPOLE
FIGURE 1-1

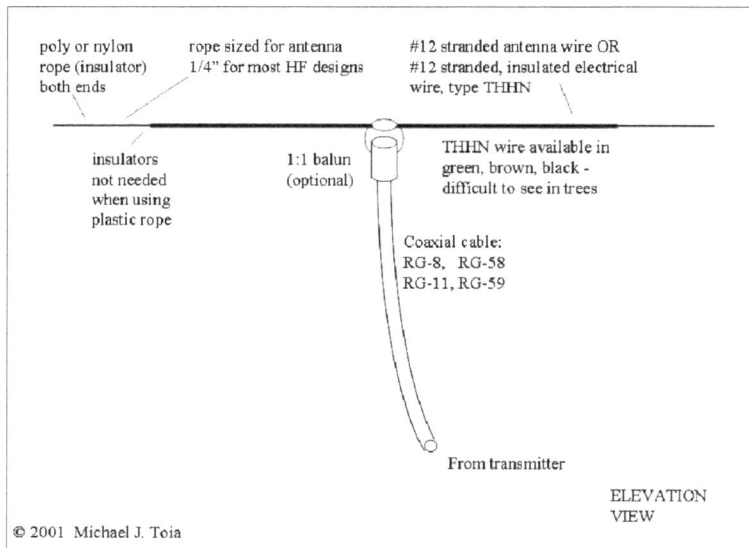

poly or nylon rope (insulator) both ends

rope sized for antenna 1/4" for most HF designs

#12 stranded antenna wire OR #12 stranded, insulated electrical wire, type THHN

insulators not needed when using plastic rope

1:1 balun (optional)

THHN wire available in green, brown, black - difficult to see in trees

Coaxial cable: RG-8, RG-58 RG-11, RG-59

From transmitter

ELEVATION VIEW

© 2001 Michael J. Toia

USE OF A BALUN TO MATCH THE DIPOLE
ENDS INSULATED WITH PLASTIC ROPE
BALUN SHOULD BE SUPPORTED WITH A LANYARD.
FIGURE 1-2

Figure 1-1 shows a typical installation for a simple half-wave dipole. For handy reference, the table shows wire lengths for operating frequencies in amateur bands from 160 meters (1.8-2 MHz) to the 23 cm band (1240-1300 MHz). The figure is an *elevation view,* or a view looking at it from the side, with the earth at the bottom, and the sky at the top.

End insulators are optional. A few feet of plastic rope (polypropylene or nylon) can be used in their place. 1/4" diameter rope sold at building supply centers and most hardware stores works well.

The wire should be at least #14 copperweld, or #12 stranded antenna wire for HF antennas. However, this material is not always handy. #12 insulated electrical wire, readily available at building supply centers and electrical supply stores, is quite convenient. Type THHN wire comes in a variety of colors, with green, brown, or black blending into a backdrop of trees so well that this antenna is difficult to see from more than 30' away. THHN is usually provided with a tough nylon outer jacket that resists abrasion when in contact with tree branches, and insulates the antenna quite well for operation with up to 150 W RF power. For better flexibility, use stranded, rather than solid, wire.

Figure 1-2 shows omission of the end insulators, and use of plastic rope for supporting the ends. Installation of the dipole requires convenient end supports - trees, a mast or two, the edge of a roof along a house or garage - to stretch out the antenna. With THHN wire, I can string my dipole ends *through* the crowns of trees with good results. My present antenna (as of this writing) has been drawn through oak, maple, and chestnut trees for more than ten years, yet still loads well and lets me work a lot of DX: I work VK's most any morning, and work Europe/Asia most any late evening on open bands - particularly 20 meters.

Figure 1-2 shows more detail on rope and wire specifications. It also shows installation of an optional 1:1 balun. The balun should be used, but is often omitted without sacrifice of good performance. If you have one, use it. If not, do without. If a balun is used, a support may be used to hold it up, as in Figure 1-5.

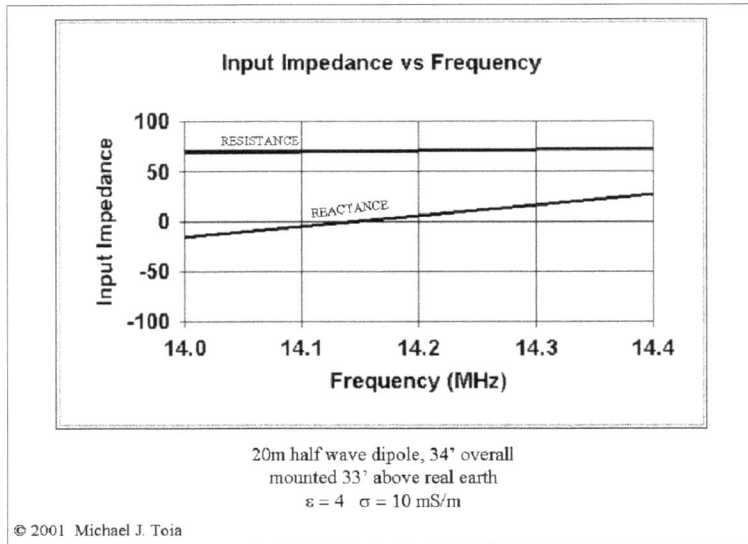

Input Impedance vs Frequency

20m half wave dipole, 34' overall
mounted 33' above real earth
$\varepsilon = 4$ $\sigma = 10$ mS/m

© 2001 Michael J. Toia

**INPUT IMPEDANCE CALCULATED WITH NEC-2
FIGURE 1-3**

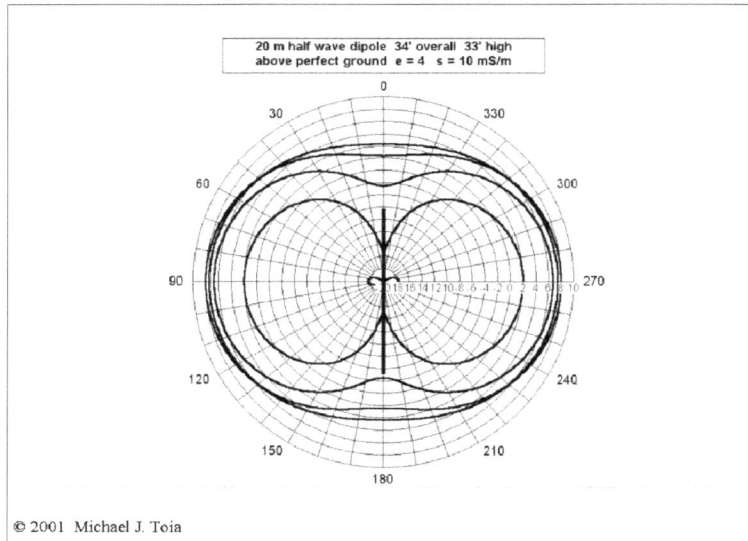

20 m half wave dipole 34' overall 33' high
above perfect ground e = 4   s = 10 mS/m

© 2001 Michael J. Toia

2 dB steps 10 dBi max
**20 METER HORIZONTAL DIPOLE
AZIMUTH GAIN PATTERNS (dBi) FOR
1, 10, 20, 30, AND 40° TAKEOFF ANGLES
FIGURE 1-4**

4

Figure 1-3 shows the input impedance of a 20 meter dipole, a half wave high, over real earth typical of U.S. soils. The soil dielectric constant is 4, and its conductivity is 10 millimhos per meter, or 10 millisiemens/m. The dipole is resonant - neither positive (inductive) nor negative (capacitive) at 14.15 MHz, and stays within ±25 ohms reactance across the band. The resistive part of impedance is seen to stay fairly constant across the band at nearly 75 ohms.

Figure 1-4 shows the calculated gain, in dB above isotropic, for the 20 meter dipole. Five plots are given, for takeoff angles of 1, 10, 20, 30, and 40 degrees, respectively, seen with increasing gain along the 0° azimuth axis. The plot for 1° is a tiny figure "8" at the center. The dipole orientation appears as a thick vertical line on this plot. Note that the antenna does not fire "off its ends" for low takeoff angles, but at 40° takeoff angle, its gain off the ends is 2.1 dBi, and broadside it is 8 dBi. This gain comes from the strong ground reflection, that adds nearly 6 dB to the antenna's free-space 2.16 dBi gain. See further explanation of this effect in Chapter 5.

Figure 1-5 shows installation of a dipole as an "inverted vee." This requires a single center support, and permits the two halves of the dipole to slope down to low supports near the ground.

When installing the inverted vee, keep the wire ends above head level by a few feet. The ends of transmitting dipoles have high RF voltages on them. At about 100 W. RF power, an arc can be drawn to the head or hand: this causes a deep RF burn that goes to the bone. It is painful and heals slowly. It also smells *awful*: the one smell worse than that of burning flesh is one's *own* burning flesh!

The inverted vee is a pretty good antenna for working DX. It combines the simplicity of a dipole with a good deal of vertically polarized radiation off its ends at a fairly low takeoff angle. [2]

---

[2] Angle between the ground and the ray leaving the antenna toward the ionosphere, for HF antennas. In general, the angle between the antenna's main lobe and the ground.

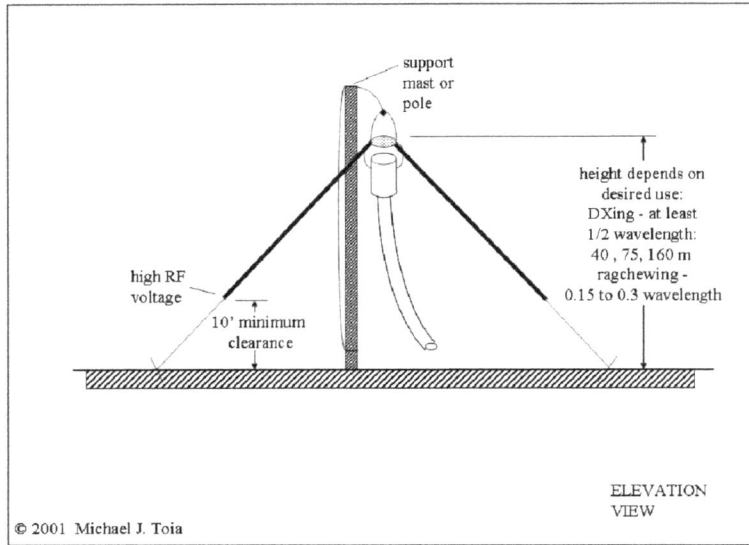

© 2001 Michael J. Toia

support
mast or
pole

height depends on
desired use:
DXing - at least
1/2 wavelength:
40 , 75, 160 m
ragchewing -
0.15 to 0.3 wavelength

high RF
voltage

10' minimum
clearance

ELEVATION
VIEW

INVERTED VEE INSTALLATION
MAY BE USED WITH OR WITHOUT BALUN.
REQUIRES ONE TALL SUPPORT AND TWO LOW SUPPORTS.
FIGURE 1-5

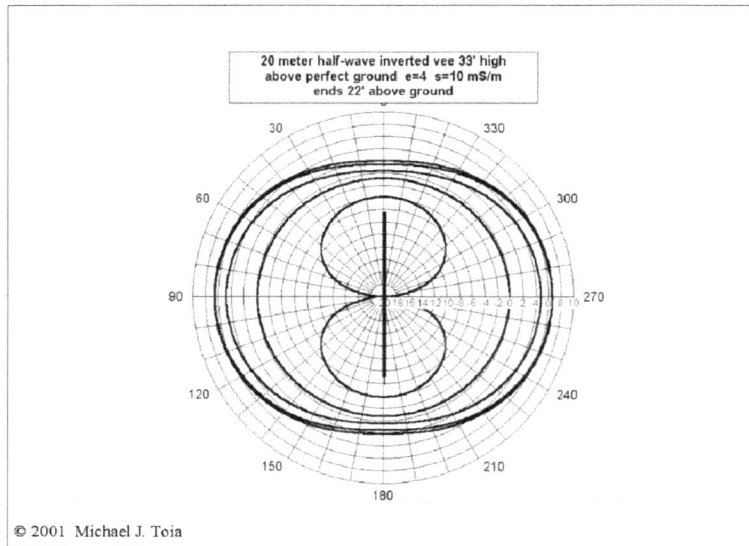

20 meter half-wave inverted vee 33' high
above perfect ground e=4 s=10 mS/m
ends 22' above ground

© 2001 Michael J. Toia

2 dB steps 10 dBi max
20 METER INVERTED VEE
AZIMUTH GAIN PATTERNS (dBi) FOR
1, 10, 20, 30, AND 40° TAKEOFF ANGLES
FIGURE 1-6

6

Note the major difference between patterns in Figure 1-6 (inverted vee) and Figure 1-4 (dipole mounted horizontally). The inverted vee has a lobe of -4 dBi gain *off the ends* at 1° takeoff angle! This lobe gets stronger for higher takeoff angles. By contrast, the maximum lobe for the horizontal dipole at 1° takeoff angle is -18 dBi. That's quite a difference. At 10° takeoff angle, the relative gains are -1 dBi, rather uniformly for the inverted vee, vs. 2 dBi broadside but a strong null off the ends for the horizontal dipole. This is a reason the inverted vee is often preferred to a straight dipole. By the way, its input impedance has dropped to very close to 50 ohms, resistive, with the usual variation in reactance across the band.

A last configuration, requiring three supports at three corners of a square, is used to achieve more omnidirectional coverage. Dipoles do not radiate well off their ends: this "quadrant" design lets one leg radiate, say, N-S while the other radiates E-W. Figure 1-7 shows some of the installation details.

The quadrant's pattern appears in Figure 1-8. Its input impedance is reactive as before, within ±25 ohms, and its resistance is very nearly 50 ohms across the band. Its main claim to utility is that it radiates a more uniform signal at all azimuth points, relatively independent of elevation angle above 10°. Installed for 75 or 40 meters, this antenna would be a good one for all round rag chewing and working a bit of DX, too.

This fairly summarizes some thoughts on the common dipole antenna, and establishes some references for expected gain and elevation angle performance.

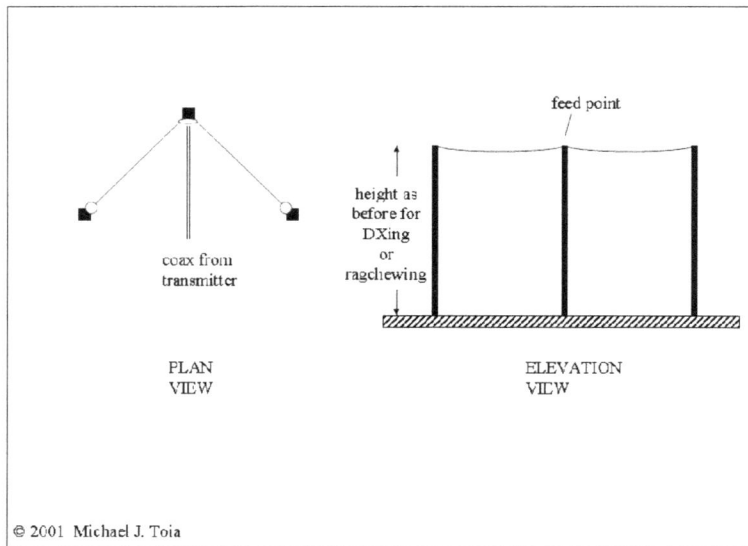

QUADRANT INSTALLATION OF DIPOLE
FOR MORE UNIFORM OMNIDIRECTIONAL COVERAGE
FIGURE 1-7

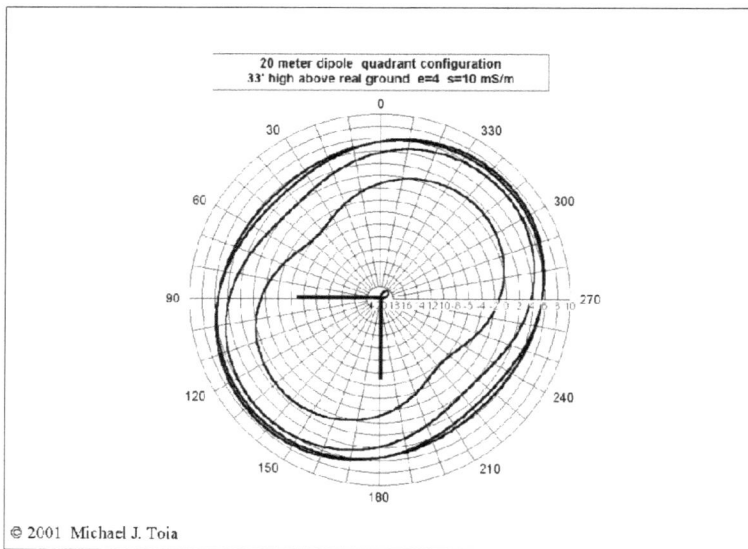

2 dB steps   10 dBi max
20 METER DIPOLE QUADRANT ANTENNA
1, 10, 20, 30, 40° TAKEOFF ANGLE PATTERNS (OUTWARD FROM CENTER)
FIGURE 1-8

NOTES

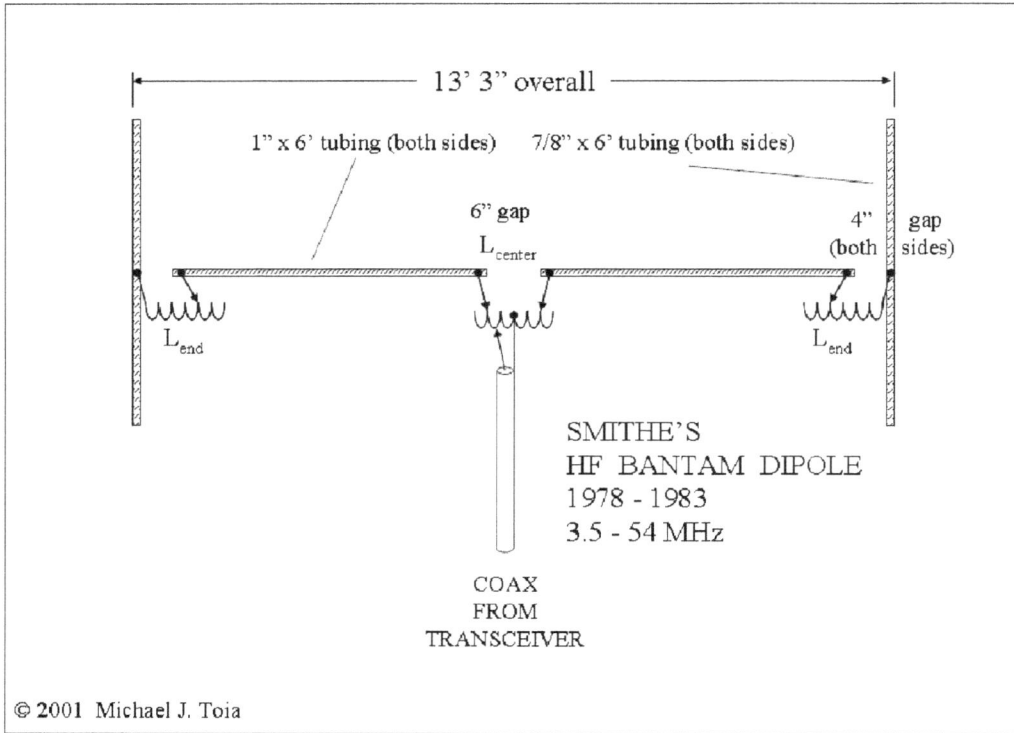

13' 3" overall

1" x 6' tubing (both sides)   7/8" x 6' tubing (both sides)

6" gap
L_{center}

4"
(both

gap
sides)

L_{end}

L_{end}

SMITHE'S
HF BANTAM DIPOLE
1978 - 1983
3.5 - 54 MHz

COAX
FROM
TRANSCEIVER

© 2001 Michael J. Toia

DESIGN DETAILS OF THE ORIGINAL HF BANTAM DIPOLE
SEE TEXT FOR COIL DETAILS
FIGURE 2-1

2

THE 'SMITHE'

HF BANTAM DIPOLE

Some time ago I was challenged to produce an indoor antenna for use on HF bands. Over the years many attempts have been made to solve this problem, and designs have offered mediocre results. The problem seems to be an attempt to use mobile whips with ground systems that are not well controlled, aggravating problems such as RFI in the shack, poor performance, and so on. I came to appreciate that the task was to make a *perfect* ground in the form of a phantom ground, between two loaded monopoles back-to-back.

Sound familiar? Yes, I had to build a dipole antenna that would be small enough to fit into an apartment while working. The result was a commercial product, sold in 1978-83 as the "Smithe HF Bantam Dipole," and was unique enough to garner U.S. Patent 4,207,574. The Bantam works on only **one frequency at a time**, and is just a short, end-loaded dipole. Its original version could be tuned to *any* frequency in the range of 3.4 to 54 MHz. Some of them were used by US Government operations on other than ham HF frequencies with good results.

Here's the basic design.

The frame of the antenna consisted of a 12' piece of 1" aluminum tubing, cut in the middle and separated to form a 6" gap. A plastic insulator held the two pieces together, allowed attachment of a small mast to support the structure, and provided a mounting for a center loading and feed coil structure.

The ends of the tubing were coupled to tee-shaped plastic end insulators. A 6' piece of 7/8" diameter aluminum tubing passed through these insulators. These two end tubes formed capacity "end hats" for the main dipole. The whole configuration looked like a rather elongated letter "H" as in Figure 2-1.

A piece of B&W "miniductor" air-wound coil stock, cut into three sections, provided the three inductors. The stock was 3" in diameter, #18 wire, 10 turns per inch, 10" long, for a total of 100 turns. It was cut to make a center inductor 3" long, trimmed to 29 total turns, and two end inductors, each 3¼" long, trimmed to 32 turns each.

The centers of the end tubes connected to the end inductors as shown, and a moveable clip connected the ends of the center tube to a tap on these coils. The ends were always adjusted to be symmetric: the same number of turns was used for each coil. The coil stock and clips may still be available from B&W. Check their web page.

The Bantam was manufactured with the two 6' center tubes cut in two, each 3' long to expedite packaging and shipping. Naturally one asks what would happen if only two, rather than four, of the 3' tubes were used, shortening this antenna to 7' 3" overall. This was, in fact, recommended in the instruction manual. The antenna no longer operates down to 3.4 MHz, but picks up at 3.8 MHz.

Typical mounting of the Bantam is about 6' - 7' above ground or the floor. Tune this antenna by attaching a coaxial cable to the center coil. Connect the shield to the center of the coil. Connect the inner ends of the center tubes to taps on the center coil, symmetrically either end of the coax shield connection, with moveable clips and flexible pigtails about 6" long. Finally, connect the coax center conductor to the center coil with a moveable clip.

Adjustment to frequency is a bit laborious, and can take a while - ten or so minutes for the first tuning, then a minute or so thereafter. First, establish "pre-tap" positions on the center coil, about 30% of the way toward the outer ends. Connect the coax to a receiver. Do a listening test with some noise source - this could be a fluorescent tube or just background noise in the house. Move the end coil taps to find a point of maximum noise into the receiver. A good technique is to tune the receiver a bit across the shortwave band to see where the antenna is approximately tuned. Then adjust the end coil taps to more inductance to lower this frequency, or less to raise it.

Bring the coax away from the center at a right angle from the Bantam to the floor or ground, then along the floor or ground for about 4' or so. Once things are close to frequency, apply a bit of transmit power, and take SWR readings at the edges and center of the band being tuned.

CAUTION

**THIS ANTENNA DEVELOPS VERY HIGH VOLTAGE
ON THE END TUBES (CAPACITY HATS)
AND CAN CAUSE SEVERE RF BURNS**

**ARCS OF 3-4" FROM THE ENDS TO COMBUSTIBLE
MATERIALS COULD IGNITE A FIRE**

**THESE EFFECTS OCCUR WITH AS LITTLE AS
50 WATTS RF INTO THE ANTENNA**

<u>DO NOT TOUCH THE ANTENNA WITH RF POWER APPLIED!</u>

Unkey the transmitter (remove RF power), and adjust the end coil taps (symmetrically) to the best position at this time, perhaps a bit on the high frequency side.

Then adjust the tube taps on the center coil to bring resonance to the desired spot, and adjust the coax center conductor tap to bring the VSWR down to a low value.

AT THAT TIME MAKE A NOTE OF ALL SETTINGS FOR THIS FREQUENCY!

This antenna has very high Q, and must be re-tuned every 50 kHz or so across 80 meters, or every 100 kHz across 40 meters. On the other bands, it is necessary to re-tune when going from the CW to the phone portion.

The original Bantam was supplied with a short mast, a ¼-20 threaded hole, and was designed to be mounted on a camera tripod. The center insulator mount had a swivel to permit either horizontal or vertical mounting, with the idea that vertical mounting would be better for DX work. But the short anecdote that follows, while true, remains a mystery to me.

I had been challenged to design this antenna for sale at a local ham radio emporium (no longer in business - but one of the best I had ever seen.) The proprietor and I were in his parking lot, listening to 20 CW with the prototype Bantam on a tripod, 7' above the blacktop surface. A UA was calling CQ. I listened, and turned the antenna vertically.

To our surprise, the signal, although fading about 2-3 S-units, seemed to drop about 2 S-units. So the proprietor turned the antenna horizontally. The signal seemed to recover about 2 S-units. "Knowing" that vertical polarization would work better, I returned the Bantam to the vertical position: the signal seemed to drop back 2 S-units. The proprietor turned it back horizontal, with signal recovery apparent.

Finally the UA finished his CQ. I was demonstrating the antenna for receive, but my friend, the proprietor, urged me to call the UA. I did. *He answered*! We had a nice, short chat from the Bantam in Maryland, to the (then) USSR.

We returned to the store, and I got my first order for a half-dozen finished antennas. Over five years about 200 or so of these antennas were sold. Only one was returned - for a really botched job of manufacture. In fact, the customer wanted a decent Bantam, which was provided. No others were ever returned, but a lot of positive mail was received.

I no longer manufacture this antenna nor provide parts. Another firm, however, has approached me for the full design and technical expertise, and may well re-introduce a new version of the Bantam. Time will tell.

NOTES

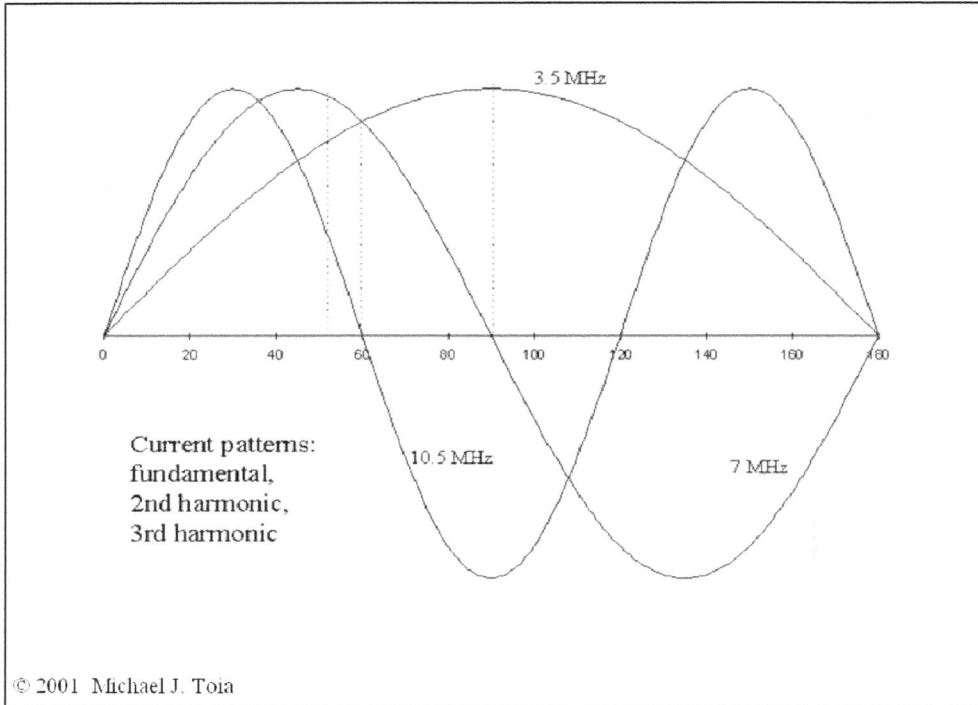

CURRENT PROFILE ON A HALF-WAVE DIPOLE
AT FUNDAMENTAL, 2ND, AND 3RD HARMONICS
FIGURE 3-1

3

THE "SMITHE" WINDOM

AND THE K3MT WINDOM

80 T0 10 METER ANTENNA

This antenna was marketed in the late 70's and early 80's as Smithe's Windom. It was designed to cover 80, 40, 20 15, and 10 meters. By serendipity, it also covered the 17 and 2 meter bands - the latter not very useful, though.

What *is* a Windom? It's a class of off-center fed dipole antennas designed to work on a fundamental frequency and some of its harmonics.

A Windom begins with a center-fed, half-wave dipole at the lowest desired operating band. Since the early HF bands were (by design) harmonically related (3.5, 7, 14, and 28 MHz), an attempt was made to operate the antenna on a fundamental frequency at the lowest band, and harmonically at higher bands. The center-fed, half-wave dipole works fairly well on all <u>odd</u> harmonics, because the center of the antenna has a current maximum at its fundamental and odd harmonics. That's why a 40 meter dipole works fairly well on 15 meters. It won't work well on 20 or 10 meters, though, because on even harmonics, the center of the antenna has a current minimum; it's a high-impedance, center-fed Zepp antenna on <u>even</u> harmonics. Figure 3-1 shows the current standing wave on a 3.5 MHz half-wave dipole, and the currents on the second and third harmonics (7 and 10.5 MHz.)

The antenna's overall length is 180° at the fundamental frequency, or slightly longer. My design is 138' overall - coincidentally mentioned as the length used by Windom himself in his original design, according to a few old articles on the subject. When fed at the center - 90° from one side - a good match to coax occurs on 3.5 MHz.

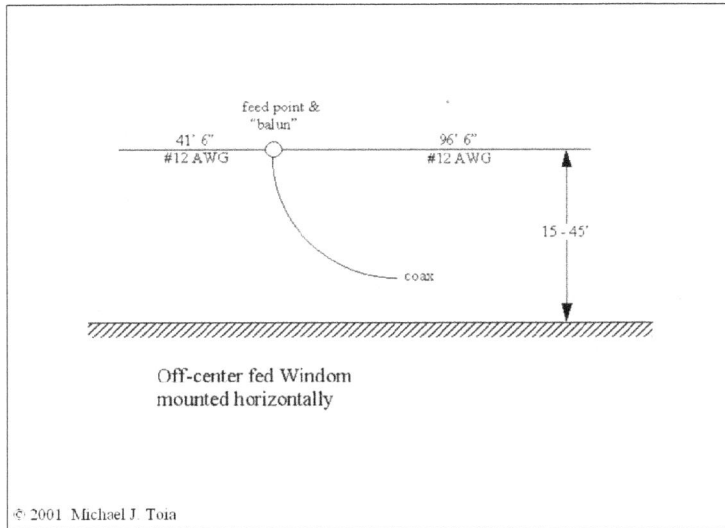

feed point &
"balun"

41' 6"
#12 AWG

96' 6"
#12 AWG

15 - 45'

coax

Off-center fed Windom
mounted horizontally

© 2001 Michael J. Toia

FIGURE 3-2

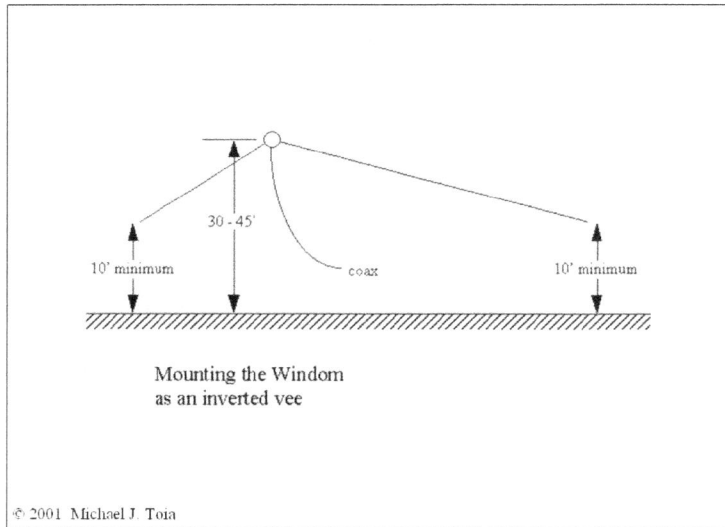

30 - 45'

10' minimum

coax

10' minimum

Mounting the Windom
as an inverted vee

© 2001 Michael J. Toia

FIGURE 3-3

But the match at 7 MHz is bad because the current is a minimum, and the impedance is very high. So feed it 60° from the left end. At 3.5 MHz the current is lower than at the center, and the voltage is higher, so the feed impedance is higher - over 100 ohms. But the antenna is still resonant, so the reactance is low! What you've done is to increase the feed point resistance.

Look now at the action on 7 MHz. The feed point is no longer at a current minimum. Therefore, the second harmonic feed impedance is quite a bit lower than it had been earlier, and is in the range of a few hundred ohms. Since the antenna is resonant here, too, it has low reactance. This design works well on 80 and 40 meters.

But now the feed impedance at 10.5 MHz is poor, because the 3rd harmonic current standing wave is now a minimum. So try feeding it at about 52 degrees from the left end. Here the match at 3.5, 7, and 10.5 MHz is fairly good. The impedance at all three is now somewhere around 200 to 400 ohms.

You can play around with this idea all day, and if you build the antenna of Figure 3-2, you will find it works well on 80, 40, 20, 17, 15, 12, and 10 meters - plus 2 meters as well, provided you pay attention to the balun! To boot, the balun matches 50 ohm coax without an antenna tuner. I admit, that this is a compromise design, and a tuner helps on the low end of 80 meters a bit, and on the high end of ten. But without a tuner, and with a fussy rig - my Drake TR-7 - a lot of DX has been worked on all bands, from 80 through 10 meters. This particular design has been installed at K3MT from 1978 to the present day. It's about the only HF antenna my daughter, KF4LGR, has used - and she's worked a good deal of Europe, Russia, Australia, a KH8, and several others - with not much time on the air!

I sometimes put a windom up more like an inverted vee, as shown in Figure 3-3, particularly when setting up for Field Day.

Balun details

What about the balun? The original unit sold with the Smithe Windoms is a Guanella-type, or current, balun, as opposed to a Ruthroff-Sevick "voltage" balun. Guanella baluns are coiled transmission lines connected in series at one end and in parallel at the other. Since the antenna design impedance was measured to be between 300 and 600 ohms, a 9:1 down-converting balun with three 150 ohm lines was used.

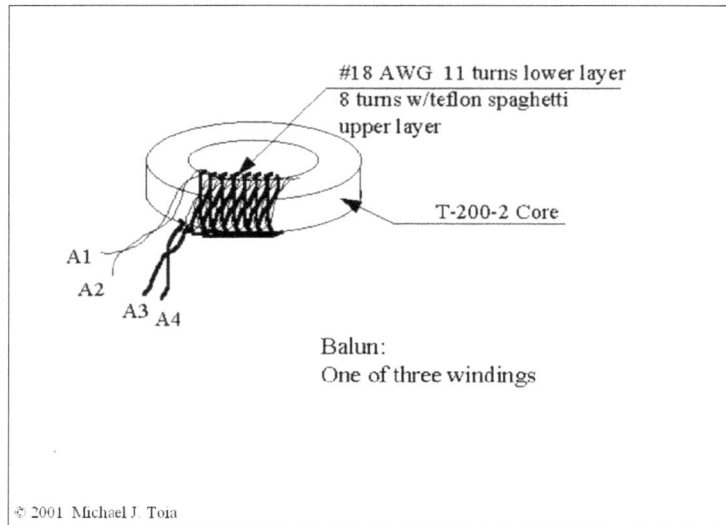

#18 AWG 11 turns lower layer
8 turns w/teflon spaghetti
upper layer

T-200-2 Core

A1
A2
A3 A4

Balun:
One of three windings

FIGURE 3-4

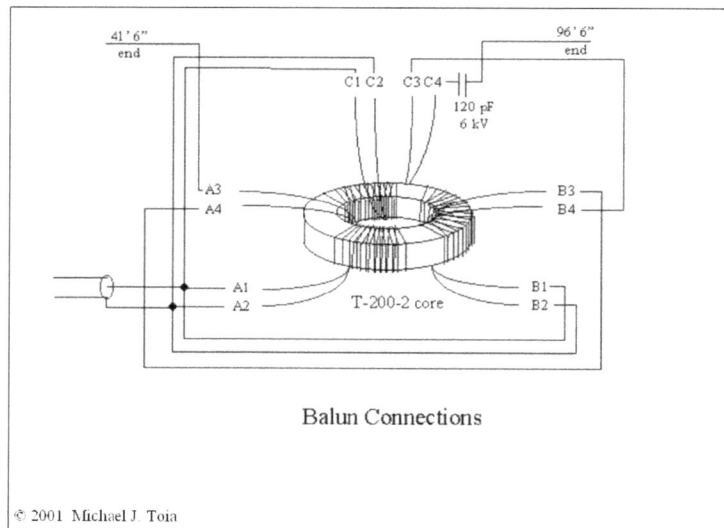

41' 6"
end

96' 6"
end

C1 C2   C3 C4

120 pF
6 kV

A3   B3
A4   B4

A1   B1
A2   B2

T-200-2 core

Balun Connections

FIGURE 3-5

To build the Guanella balun, obtain an Amidon T-200-2 core, tape it with three layers of black poly electrical tape, and obtain some #18 AWG magnet wire with a bit of #17 AWG teflon spaghetti. Twist the magnet wire to make three twisted pairs - about one twist per inch. Wind 11 turns of one pair on the core, and slip the teflon spaghetti over each lead of the remainder (untwist it a bit to do this.) You should have constructed a twisted pair, with two pieces of teflon over each wire of the last several inches. Then wind 8 more turns back overtop the 11 turn winding, to make a total of 19 turns. Do this with the other two twisted pair lines as well. Space them on the core so no two lines overlap.

Figure 3-4 shows a single winding on the core. Make two more windings like it.

Get an ohmmeter to check continuity. Label the lines A, B, and C, and their ends 1 & 2 where the uninsulated wire starts onto the core, and 3 & 4 where the wire (insulated with the spaghetti) leaves the winding. Pay attention to the wiring detail that follows, and use your ohmmeter to check your work. Label the wires so there is continuity from:

* A1 to A3 (opposite ends of the same wire)
* A2 to A4
* B1 to B3
* B2 to B4
* C1 to C3
* C2 to C4

Refer to Figure 3-5 as a guide. Wire the balun as follows:

* Connect A1, B1, and C1 together. These will connect to the center conductor of the coax.
* Connect A2, B2, and C2 together. These must connect to the coax braid.
* Connect A3 to the **short** end of the windom. This is important!
* Connect A4 to B3, and B4 to C3.
* Connect C4 to a 110 pF, 6 kV capacitor.
* Connect the other end of this capacitor to the **long** end of the windom.

The short end of the windom is 41' 6" long, usually made of #12 electrical or antenna wire. The long end is 96' 6" long. I prefer to use #12, type THHN insulated electrical wire - it can be threaded through or over trees, and works well with about 100W of RF into the antenna without arcing.

You are now ready to install and enjoy your windom. If you have the same luck that K3MT and daughter KF4LGR have, it will have been worth all the trouble!

A Re-design

Since designing the first "Smithe Windom," I've investigated the antenna a bit further. I now find I can extend the range to all HF bands except 30 meters by re-designing the balun. The new balun is a Ruthroff-Sevick unit, consisting of a bifilar winding of two, parallel, #18, formvar insulated magnet wires on an Amidon T-200-2 core as above. It takes about 8 feet of wire to make the winding. Keep the two #18 wires parallel without twisting, and wind them fairly evenly about the core - there will be a space of about an eighth of an inch between the paired windings.

As before, label the wires A1 and A2, B1 and B2, so that A1 and A2 are the ends of the same wire: ditto for B1 and B2. Then connect A2 to B1. Connect A1 to the short end of the antenna, and B2 to the long end. **Do not install the series capacitor** in this design. Connect the coax shield to the junction of A2/B1, and its center conductor to A1 and the short end of the antenna.

So built, there will be some compromise with this antenna. It does not work on 2 meters. It has a higher VSWR across all of 20 meters: although it can be operated well with coax feed directly to the transceiver, a tuner definitely helps get the maximum out of solid state rigs on that band. It also benefits considerably from a tuner on the low end of 80 meters.

But the advantages gained are operation on 80, 75, 40, 20, 17, 15, 12, and 10 meter bands, with improved performance on the high end of 10 meters. It can also be operated on 30 meters, but a tuner is required to work that band - the coax VSWR can be up to 5:1.

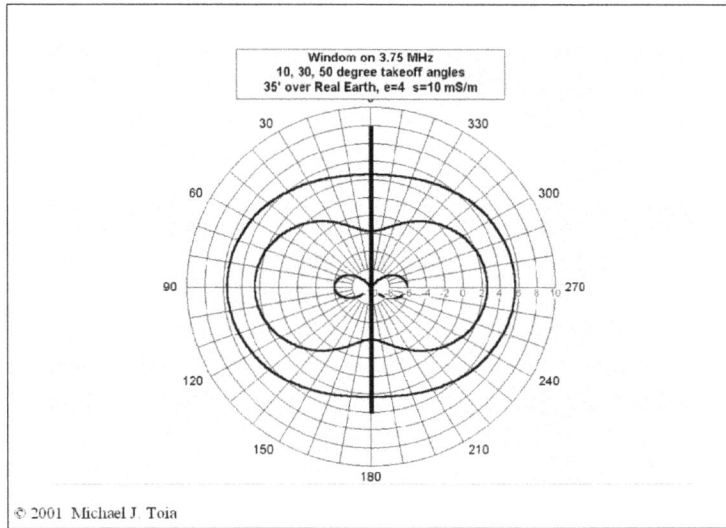

2 dB steps   10 dBi max
80/75 METER PATTERN
FIGURE 3-6

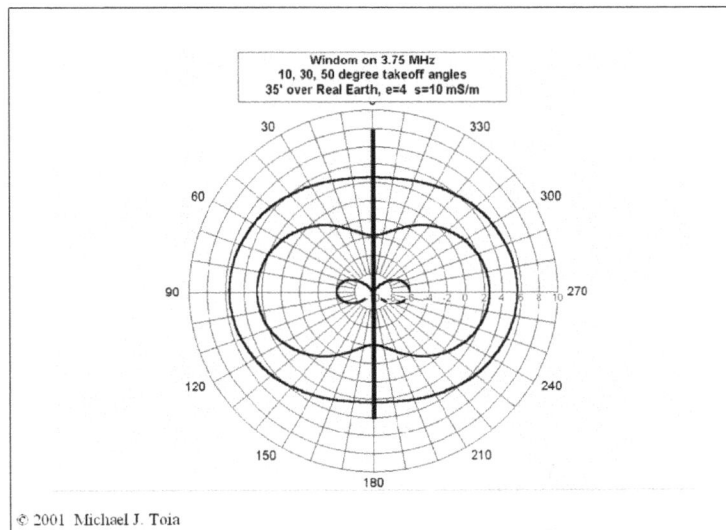

2 dB steps   10 dBi max
40 METER PATTERN
FIGURE 3-7

21

In these pattern plots, the windom is aligned along the 0° - 180° azimuth line (straight up and down) and is fed slightly below the center of the plot. Figure 3-6, the 80 meter pattern, shows a typical dipole pattern for low dipoles at 80 meters - excellent for ragchewing. There is low gain at the 10° takeoff angle, but about 5.5 dBi gain broadside to the antenna at 50° takeoff angle.

Figure 3-7, the 40 meter pattern, shows the null broadside to the antenna, characteristic of two dipoles fed out of phase. It does show, however, best performance *off the ends* (along 0° and 180° azimuth) for the higher takeoff angles. At 10° takeoff angle, a four-lobed pattern results, with low gain. This is due, of course, to the fact that the antenna is a quarter wavelength high at this band - it tends to fire "straight up."

Figure 3-8, the 20 meter pattern, shows significant performance for a takeoff angle of 10° (the inner pattern) where a 4 dBi gain is found at an azimuth of ± 34°, and the back azimuths of these.

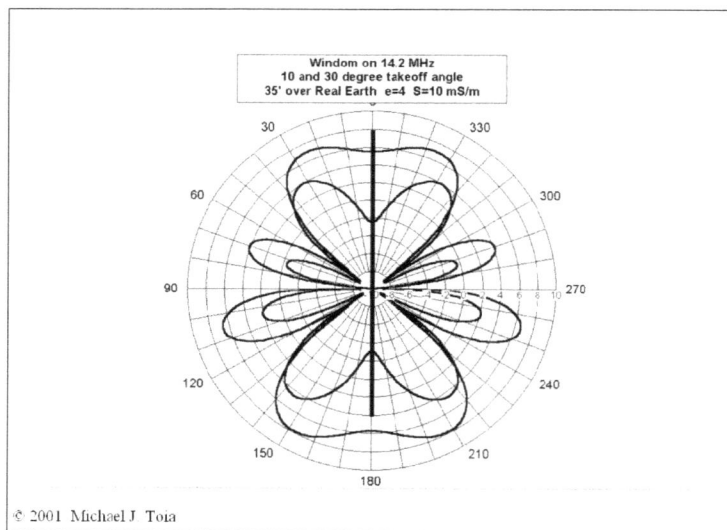

2 dB steps 10 dBi max
20 METER PATTERN
LOWER GAIN FOR 10° TAKEOFF ANGLE
OTHER PATTERN FOR 30° TAKEOFF ANGLE
FIGURE 3-8

The 50° takeoff angle pattern is not shown, but is inside (less gain than) the 10° takeoff pattern. The pattern also shows a null broadside, but good endfire gain at 30° takeoff angle off the ends.

Patterns for 17, 15, 12, and 10 meters follow along similar lines, with more lobes and slightly higher gains at the higher frequencies. The main lobes fall a bit closer to the wire, and all have nulls at right angles to the wire. Hopefully this gives you a smattering of an appreciation for the windom.

The high takeoff angle at 80 and 40 meters is largely a consequence of the antenna's height above ground. To make this antenna a good one for low band DXing, its installed height should be raised to about a half wavelength at the band to be used. See chapter five where I discuss antenna patterns at different antenna heights.

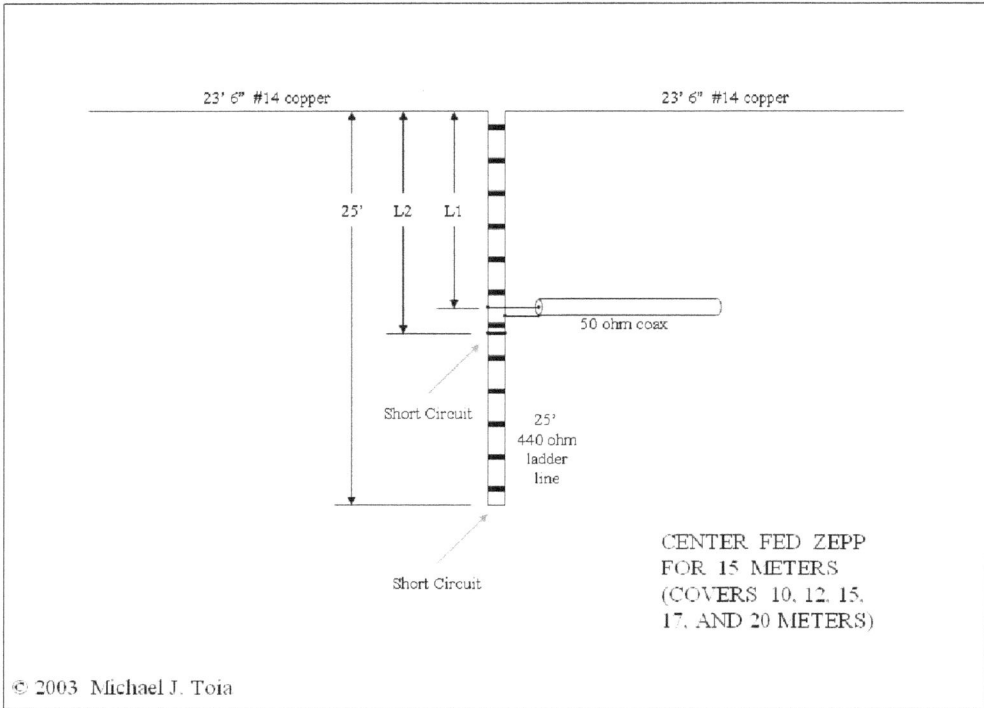

23' 6" #14 copper   23' 6" #14 copper

25'   L2   L1

50 ohm coax

Short Circuit

25'
440 ohm
ladder
line

Short Circuit

CENTER FED ZEPP
FOR 15 METERS
(COVERS 10, 12, 15,
17, AND 20 METERS)

© 2003 Michael J. Toia

STANDARD CENTER-FED ZEPP
FIGURE 4-1

- 4 -

ZEPPS, J-POLES, AND

THE LooooNG WIRE

AT W3NKI

A Note on Antenna Matching

Most amateur transmitters need a 50-ohm load at their antenna terminal. Antennas are often designed to present this impedance, or an antenna tuner converts the antenna's impedance to 50 ohms. Tuners are almost always mis-used. To do the job properly, they should be located at the antenna end of the coax, not at the transmitter. Tuners are also an added, and often unnecessary, expense.

Any antenna can be tuned to 50 ohms with two transmission lines. The first is in series, between the antenna and the 50 ohm line. The second is connected in parallel across the 50 ohm line. Look at Figure 4-1. The line between the antenna and coax is the series transmission line. The additional line between the coax and the short is the second line. The remainder of the ladder line is unimportant [3] because of the shorting bar. By choosing the length of these two lines properly, the antenna can be brought to a 50 ohm (or another) impedance. Antennas in this chapter use this tuning method. An alternative to the two transmission line method is to run the ladder line to an antenna tuner, which then allows use of the antenna on many frequencies. I stick to the two-transmission line method here.

---

[3] In some cases, the excess transmission line may be resonant and cause trouble. It can be removed, as it serves no useful purpose other than allowing room for moving the shorting bar.

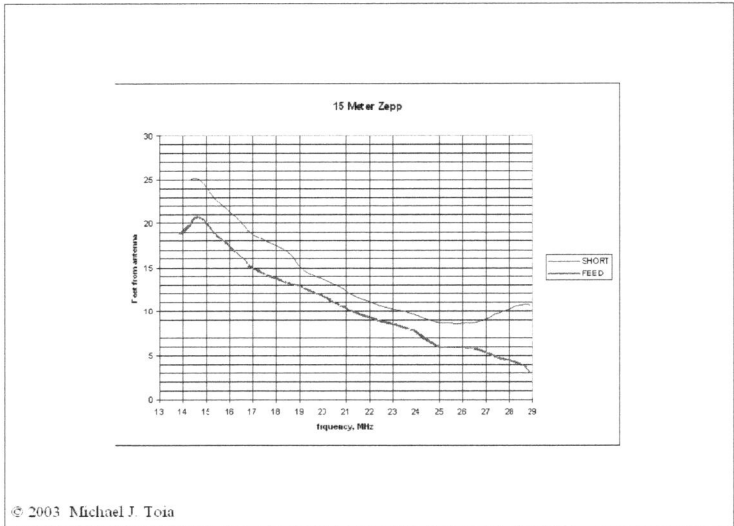

15 METER CENTER-FED ZEPP
MEASURED VALUES OF L₁ AND L₂
FIGURE 4-2

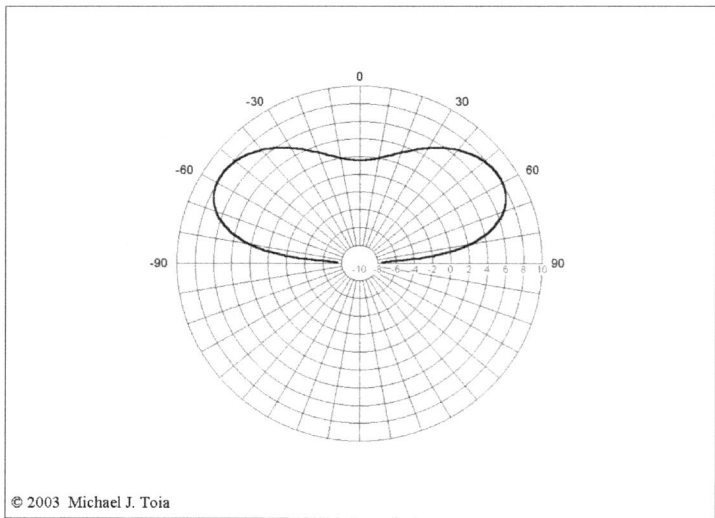

2 dB steps 10 dBi max
ELEVATION PATTERN, 15 M CENTER FED ZEPP
FIGURE 4-3

26

A Standard Center-Fed Zepp

Let's modify a half-wave dipole antenna. Look at a dipole exactly twice as long, with an end-to-end dimension of one full wavelength. It's still cut at the center and fed there. The feed impedance is now much higher than the nominal 76 ohms. NEC-4 calculations predict a feed impedance of over 2500 ohms. Figure 4-1 is an example of such an antenna for 15 meters. Its dimension is 47 feet, end to end, cut in the center for attachment of a feed line. This is called by a variety of other names, including "two half-waves in phase." It consists of two half-wave dipoles on a single line, side by side, each fed at its end where the voltage is high and the current low.

I made up some 440 ohm "ladder line" with #18 wire spaced 7/8" apart. The line was 25 feet long, and shorted at one end. The other end was attached to the two halves of the dipole, as shown. I contemplated connecting 50 ohm coax to the ladder line at distance L_1 from the antenna, and a short circuit across the line a bit farther along, at distance L_2. NEC-4 calculations predict a match at 21.2 MHz. It also predicts this antenna capable of operating on 14.1 to 30 MHz. Here are its estimates:

L_1, feet	L_2, feet	Band	SWR	2:1 Bandwidth
23.2	25.0	20 m	1.7:1	whole band
15.0	17.5	17 m	1.5:1	whole band
11.4	19.4	15 m	1.3:1	whole band
7.8	15.8	12 m	1.3:1	whole band
4.0	10.5	10 m	1.3:1	800 kHz

I built the antenna, mounted its center about 20', and its two ends about 15', above ground. A MFJ SWR analyzer showed values of L_1 (feed point) and L_2 (location of short) as plotted in Figure 4-2. While not matching the NEC-4 predictions exactly, the result is an antenna that can work the top 5 HF bands. That is, it can be tuned over a one octave (2:1 ratio) frequency range.

NEC-2 produced the antenna pattern of this Center Fed Zepp shown in Figure 4-3. Note the gain broadside to the antenna!

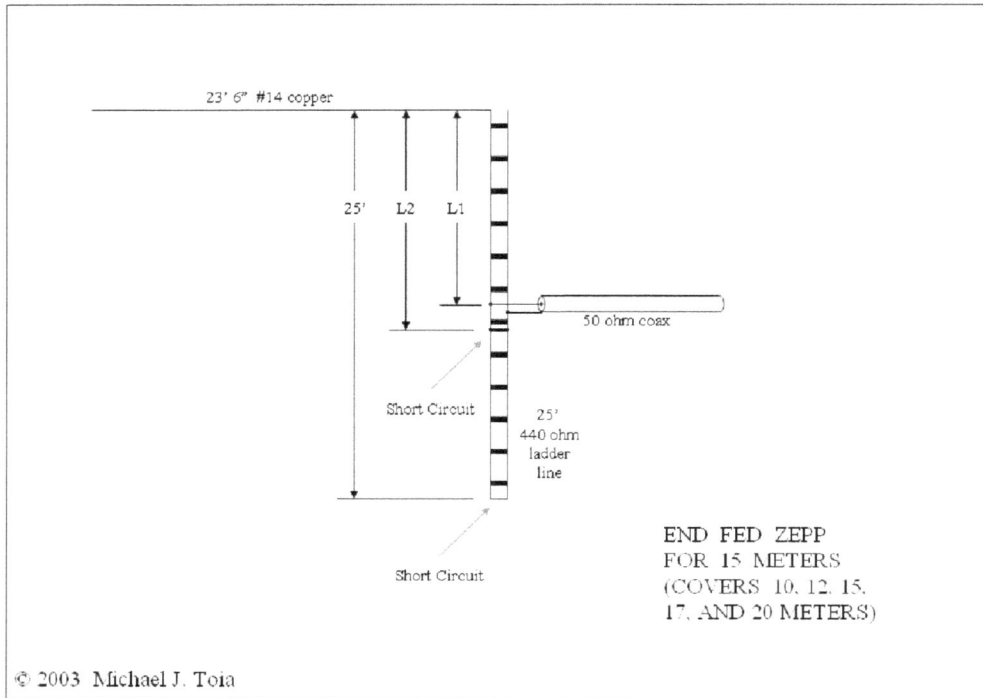

23' 6" #14 copper

25'   L2   L1

50 ohm coax

Short Circuit

25'
440 ohm
ladder
line

Short Circuit

END FED ZEPP
FOR 15 METERS
(COVERS 10, 12, 15,
17, AND 20 METERS)

© 2003 Michael J. Toia

STANDARD END-FED ZEPP
FIGURE 4-4

I tried this antenna on 20 meters, but the coax was somewhat "hot," meaning some RF was "coming down the shield" [4] into the shack. A 1:1 balun between the coax and the antenna, at the point of attachment to the open wire line, would help solve this problem.

The End-Fed Zepp

Remove one-half of the center-fed Zepp to make an *end-fed Zepp* antenna, as in Figure 4-4. NEC-4 estimated parameters for this antenna are:

L_1, feet	L_2, feet	Band	SWR	2:1 Bandwidth
24	25	20 m	1.2:1	whole band
15	16	17 m	1.2:1	whole band
11	12	15 m	1.4:1	whole band
8	9	12 m	1.3:1	whole band
4	5	10 m	1.5:1	450 kHz

Again this antenna shows promise as a five-band performer. Just modify the center fed Zepp by removing one of its halves and re-installed it as an end fed Zepp. This antenna doesn't work well on 20 meters. Mine put just too much RF into the shack. As I touched the coax or the transceiver, the SWR meter would dance around. The antenna has too much imbalance to work well on other than 15 meters. A 1:1 balun at the feed point will help somewhat. But I recommend this antenna for a single band, or bands harmonically related, such as 80/40/20 meters, or similar combination.

Extended Zepps

The center-fed Zepp above uses a pair of half-wave long wires, and the end-fed design eliminates one of them. This antenna can be "extended" by making the wire multiples of a half-wave long. Dimensions of the tuning line remain about as shown above, with some minor modification.

---

[4] More correctly, the coax was operating in its normal *differential* mode with equal, but opposite, currents on the shield and center conductor, and had some *common* mode current due to its attachment at the antenna.

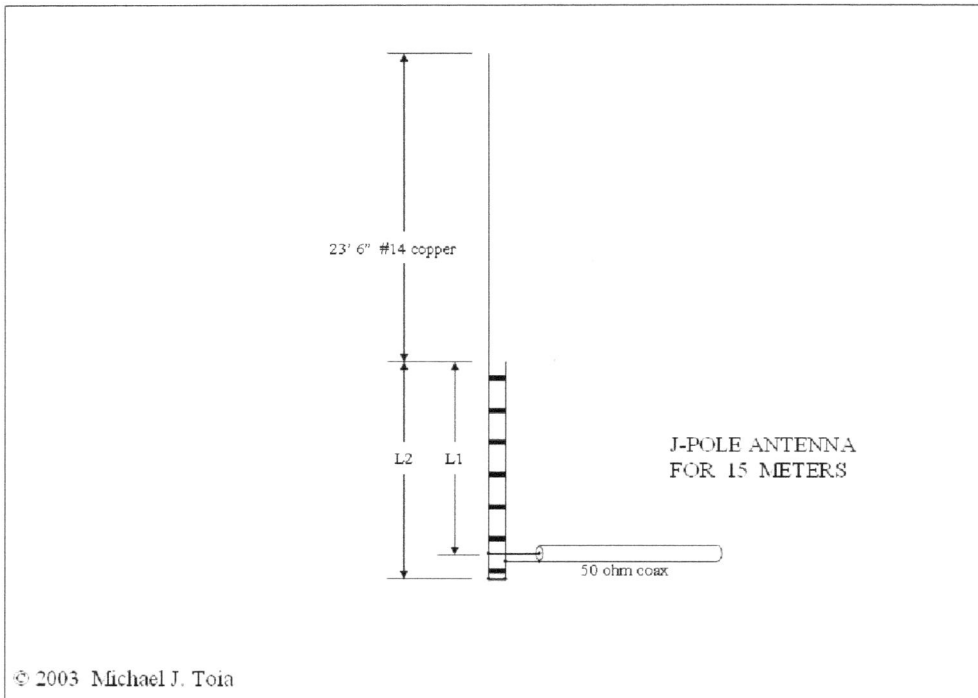

23' 6" #14 copper

L2 L1

J-POLE ANTENNA
FOR 15 METERS

50 ohm coax

© 2003 Michael J. Toia

THE J-POLE ANTENNA
15 METER VERSION
FIGURE 4-5

Extended Zepps can have significant gain, a characteristic of all long-wire antennas. See W3NKI's antenna (below).

The J-Pole Antenna

Look at Figure 4-4 a bit more closely. If I simply turn the half-wave antenna wire vertically, a J-pole antenna results. See Figure 4-5. This antenna is so-named because it is shaped somewhat like a skinny letter "J." It's often fed with balanced transmission line, such as 300 ohm TV-style twin lead, or open-wire line made of two wires spaced apart with insulators about an inch to six inches long, known as "ladder line." It can be coax fed as well.

Normal construction of the J-Pole calls for one-half wavelength of wire above the ladder line, and one quarter wavelength of line, shorted at the bottom. By bringing the feed line up the quarter-wave line a little bit, a point can be found where an excellent match results. In fact, a match for 50 or 75 ohm coax, or 300 ohm twin lead can be found for this antenna.

I built a J-pole of #14 solid, insulated, THHN copper wire for 20 meters. The top part of the antenna is 35' long, run essentially vertical. Mine was supported by a halyard thrown over a tall oak tree. The antenna wire was simply pulled up into the tree's crown. The ladder line was made of #14 copper, spaced 7/8" apart. An excellent 50-ohm match results with $L_1 = 15'$. $L_2 = 15' 6"$.

J-poles are easy and fun to build, and work quite well as a monoband antenna. They are almost always mounted with the wire vertical, but this is not a hard requirement. "Extending" a vertical J-pole will cause its pattern to raise above the ground. Although it has gain, the main lobe is not in a useful direction. [5] Radiation at 20 meters and higher frequencies that is too sharply incident on the ionosphere will not be reflected back to earth.

My 20 meter J-pole worked quite well. I've been able to work a reasonable amount of DX in the few weeks since it was put into operation.

---

[5] This is a general rule, which can have exceptions I will not consider here.

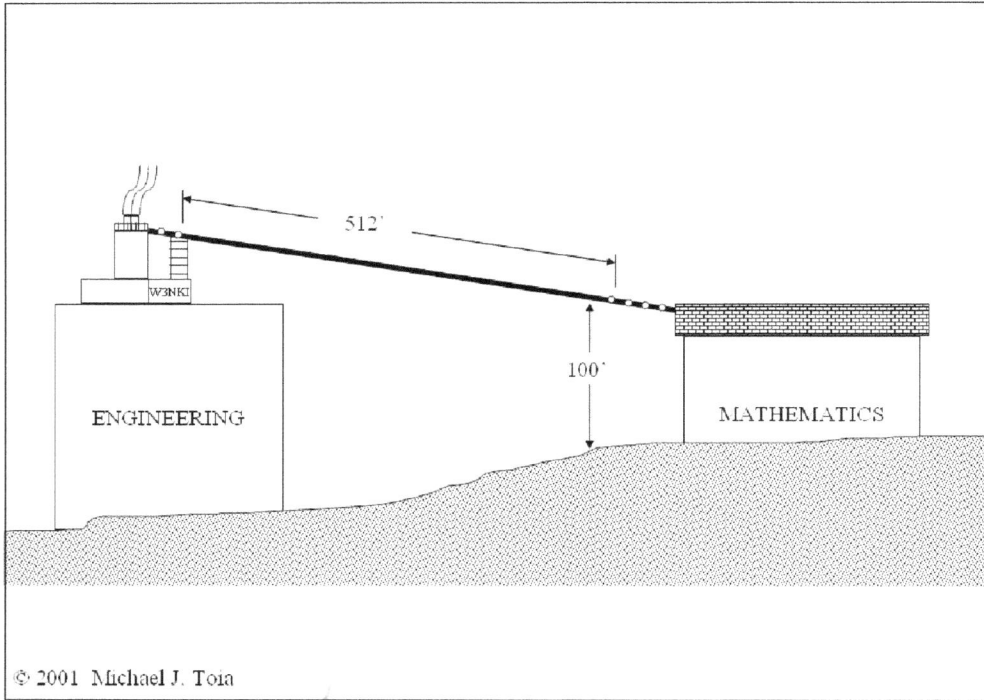

LAYOUT OF END-FED ZEPP
W3NKI - 1955
FIGURE 4-6

W3NKI's Looong Wire

Let me drift a bit in history, to 1955. I enrolled at Carnegie Tech (now Carnegie Mellon University) as a freshman physics major. Even before classes began I had joined the amateur club, (then) W3NKI. The shack was on the 4th floor of the mechanical engineering building (since renamed), under the smokestack for the central heating plant.

From the top of the stack a heavy copperweld wire ran across the campus, fastened to the eaves of the mathematics building. The wire was reportedly 512 feet long, end fed, with 600 ohm open wire ladder line: an "extended" end fed Zepp. At its low end the wire was about 100 feet above the campus. Figure 4-6 shows it schematically.

This wire was a full wavelength long on 160 meters, two wavelengths on 80, and so on. At all amateur bands it has about a 600 ohm input impedance, and it is approximately resonant on all bands. Long wires radiate enough energy with the initial surge of electrons trying to reach the end, and radiate additionally as that surge rebounds from the end, that they have "radiation dampening" (they emit radio energy - thankfully!) and tend to have low Q, and wider than normal bandwidth.

We fed the wire via its ladder line from a tuner in the shack. The transmitter was a home-brew CW rig, running an 829-B final amplifier tube. Output power was about 50 watts or so. But with that rig and the long wire, we regularly worked the world. One entire fall semester I would report to the shack about 7 AM for a daily sked with VK4VK, in Tweed Heads, Australia. We got to be close friends.

The wire was judiciously oriented along about a 60° azimuth. It had four main lobes. One covered Europe. A second covered Africa. A third covered the North Pacific. And practically all of them covered Australia.

The characteristics of these very long wires, mounted a wavelength or better over the earth, are severalfold:

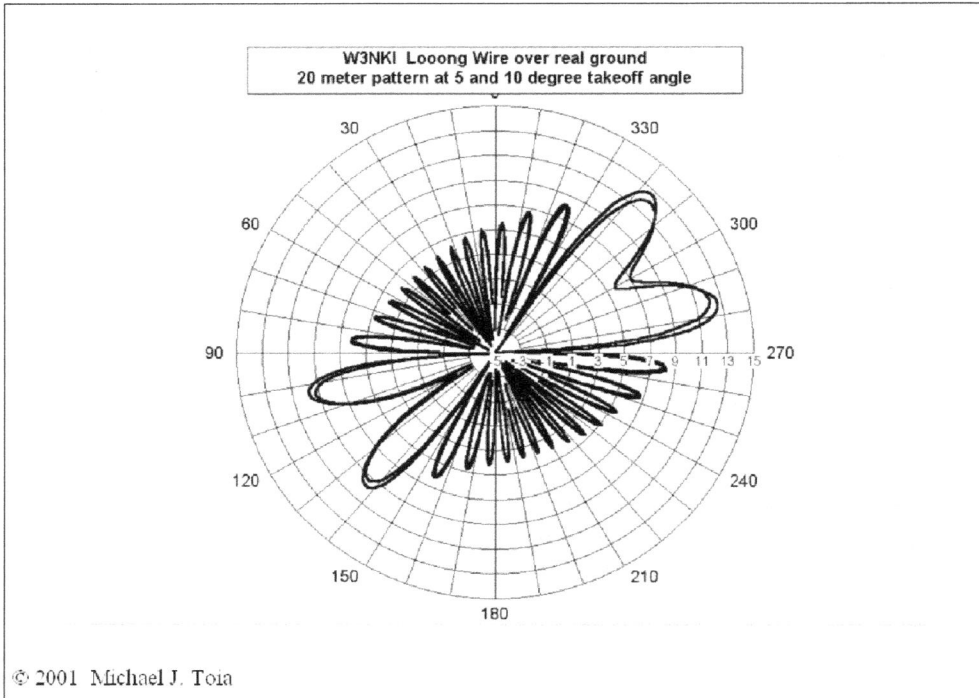

W3NKI Looong Wire over real ground
20 meter pattern at 5 and 10 degree takeoff angle

2 dB steps 15 dBi max
THE LONG WIRE ON 20 METERS
HIGHER GAIN AT 5° TAKEOFF ANGLE
LOWER GAIN AT 10°
FIGURE 4-7

© 2001 Michael J. Toia

- They have main lobes that make a small acute angle to the wire. In this case, the main lobes are at an azimuth of 44° and 76°, for a wire oriented along an azimuth of 60°.
- They can have high gain. This one had 12.5 dBi in the "forward" direction, and 9.5 dBi on the back lobes
- They achieve very low takeoff angles when high above ground.
- The main lobes have higher gain, and are closer to the wire, for the higher frequencies.

I've modeled the wire using NEC-2 to get an idea of the patterns. I had to "cheat" a bit, as that program doesn't like to handle transmission lines with one end "floating." I simply fed the antenna a quarter wave further along the wire on each band modeled. Figure 4-7 shows the estimated patterns. By accident, the angle scale is backward: East is at 270 degrees, and West, 90 degrees.

All things considered, an antenna as simple as this makes an excellent fixed-direction beam. It's the first step into design of vee-beams (two long wires making an acute angle to each other) and rhombics (two vee beams back to back) which I will not discuss in this book - the interested reader can find out a lot about long wires, vee beams and rhombics in Jasik's very interesting *Antenna Engineering Handbook*.

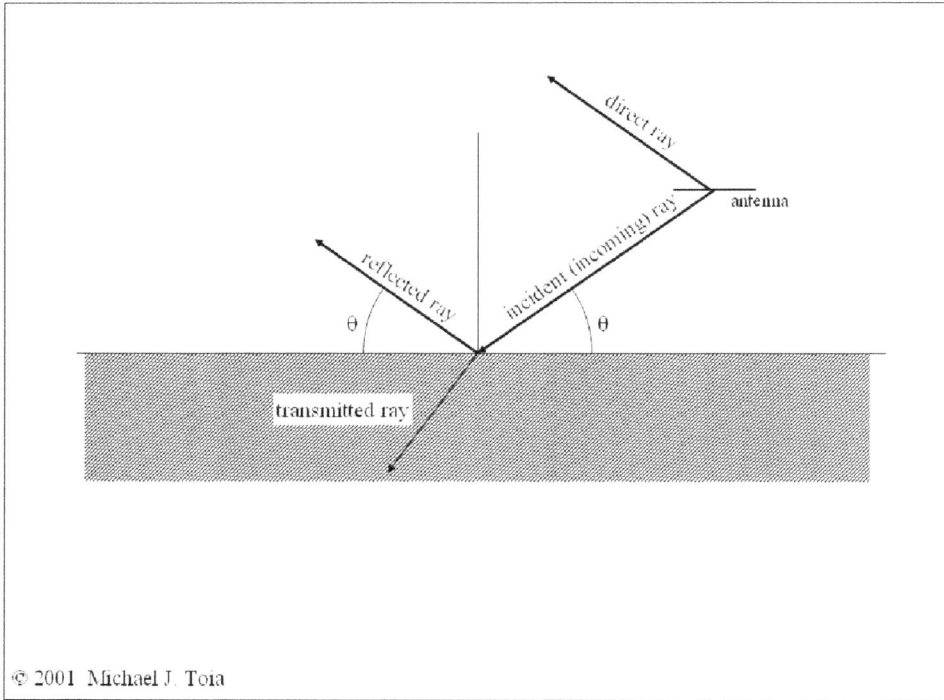

RADIO RAY STRIKING THE EARTH
FIGURE 5-1

5

GROUND REFLECTIONS:

WHY HORIZONTAL ANTENNAS

OUTPERFORM VERTICALS

Antenna theorists often visualize "rays" of radio energy radiating like straight lines from the antenna. After all, radio works because a small group of transmitted rays hits a receiving antenna, where they push the electrons and cause a current to flow. The result, a weak current, can then be sent to a receiver where the radio signals are recovered.

When a ray hits the ground, it reacts in two ways. Part of it penetrates into the ground as a *transmitted* ray, and the remainder reflects from the ground as if it were a mirror: *the angle of reflection is equal to the angle of incidence*. That is, the reflected ray leaves the earth at the same angle that the incident ray hits. See Figure 5-1.

The transmitted ray entering the ground has no further use to us. Although there have been a number of studies and tests to use this ray for communicating with underground transmitters, such as for miners trapped in a cave-in, this ray is rapidly attenuated and become too weak for much distance into the dirt. The reflected ray, on the other hand, can add to, or subtract from, the direct ray, and change an antenna's pattern. This change is quite strong. Ground reflections have a lot to do with patterns of antennas located within a wavelength or so of the ground, such as is common for amateur radio at HF.

We calculate the strength of the reflected ray with a *reflection coefficient*, which is the ratio of the radio field strength in the reflected ray, divided by the radio field strength in the incident ray. Both field strengths are taken just at the point of reflection. The reflection coefficient has both a magnitude, from zero to one, and *also has a phase shift*, which can vary from zero to 360 degrees.

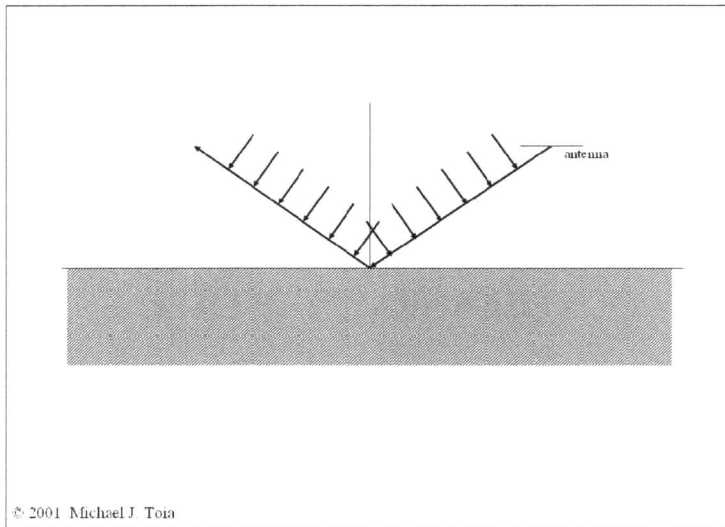

"PLANAR," OR "VERTICAL" POLARIZATION
FIGURE 5-2

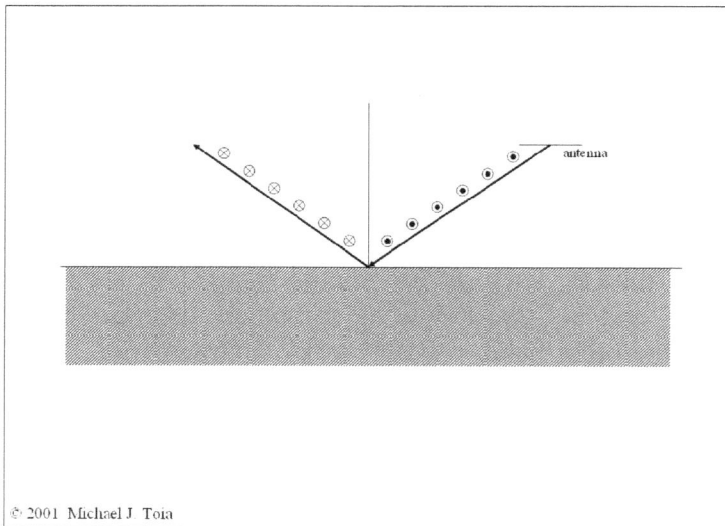

"PERPENDICULAR," OR "NORMAL" POLARIZATION
HORIZONTAL POLARIZATION
FIGURE 5-3

Theory shows that there are two distinct and different cases of reflection. One happens when the electric field (the polarization) is *in the plane of incidence*, as shown in Figure 5-2. Radio amateurs call this "vertical polarization," even though the electric field is not perpendicular to the earth (it "leans forward" a bit.) I call it *planar* polarization, since the electric field, incoming and reflected ray are all in the same plane.

The second case occurs when the electric field is perpendicular to the plane of incidence, shown in Figure 5-3. Here the electric field vectors in the incident ray point directly out of the paper at you: the circles with dots represent the tips of the arrows. After reflection, the vectors are shifted about 180° in phase, so they point away from you, into the paper - shown by the circles with "x" marks, as if looking at the south end of a northbound arrow. This is pure horizontal polarization. The electric field is always parallel to the earth. Many texts call this *perpendicular* polarization, referring to the direction of electric field. I prefer the term *normal* (at right angles to) polarization.

Any arbitrary polarization can always be made up as the sum of planar and normal polarizations. So a study of reflection coefficients for the two separate cases tells the whole story of what happens when a radio wave reflects from the earth.

Using "average" earth, with dielectric constant, $\varepsilon = 4$, and conductivity $\sigma = 10$ mS/m (mmho/m), I've calculated the reflection coefficient for both polarizations at 4 MHz. Figure 5-4 shows the magnitude of the coefficients and figure 5-5 shows their phase angles.

First look at the curves for low takeoff angle (about ten degrees) and horizontal (normal) polarization. The magnitude of the reflected ray is about 96% of the direct ray's magnitude, and its phase angle is about 178°. If the ray were at 100% of strength and exactly 180° in phase, it would be just as strong as the direct wave and, for antennas very close to the earth, would exactly cancel the direct ray: the antenna would radiate NO horizontally polarized signal at all. At reasonable heights, though, the reflected ray travels further and changes phase, so at just the right height, the two can reinforce each other, to yield a nearly 6 dB increase in radiated power.

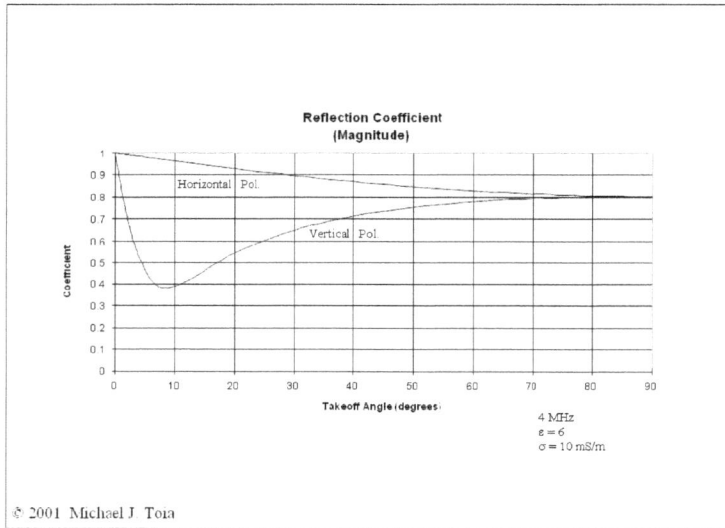

REFLECTION COEFFICIENTS (MAGNITUDE)
FIGURE 5-4

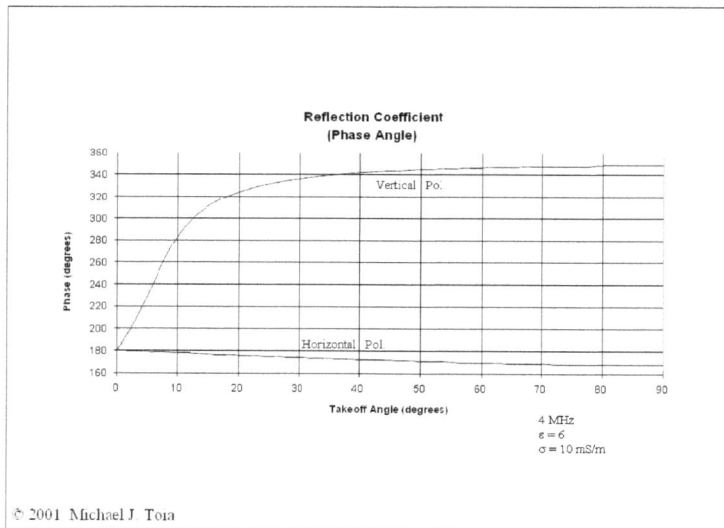

REFLECTION COEFFICIENTS (PHASE ANGLE)
FIGURE 5-5

But now look at the case for vertical (planar) polarization at 10° takeoff angle. The magnitude of the reflection coefficient is about 40% of the direct ray: it is close to a minimum value at what is called the *Brewster angle* familiar to people working in optics. Furthermore, its phase is about 285°: not only is it not strong enough to cancel the direct wave, it is about 90° out of phase with it! The net result is that, for antennas located very close to the ground, the reflected ray is quite weak at takeoff angles between 5° and 25°, so the antenna acts as if it is in outer space, with no ground near it!

Think of what this implies. It seems that a vertical antenna near the ground will radiate, but a horizontal one won't. Therefore verticals will outperform horizontal antennas. *Exactly the opposite is true*! If the antenna is mounted high enough - say, a half wave to a wavelength - the reflected ray travels a half-wavelength farther than the direct ray to reach the ionosphere, for some fairly low takeoff angle. The two add together, almost doubling the electric field, with a net increase in radiated signal strength of a bit less than 6 dB. *That's* why horizontally polarized antennas mounted at these heights are superior for working DX, compared to verticals mounted close to the ground.

At a low takeoff angle, a vertical's reflection coefficient becomes better than 85%, and its phase becomes nearly 180°, so it, too, can benefit from ground reflected gain if mounted high enough. But the required height is often out of reach of most radio amateurs on HF.

We usually put up a vertical close to (or mounted on) the ground and take our 5 dB loss in stride, for the convenience of having a simple antenna. But the attenuation of the reflected ray by the Brewster angle at low takeoff angles is interesting: I'll cover more of that in the chapters on *Beverage Antennas* and the *Grasswire*. More recent papers, unaware of my earlier publication on these antennas, refer to the same approach as *BOG* Antennas – Beverage on Ground.

Wave Tilt

When a radio ray travels from air into a dielectric, its direction changes. Except in the theory of metamaterials which I do not discuss, the dielectric constant increases: the ray entering the dielectric is more perpendicular to the surface than that in the air. Refer to Figure 5.1 to see the effect.

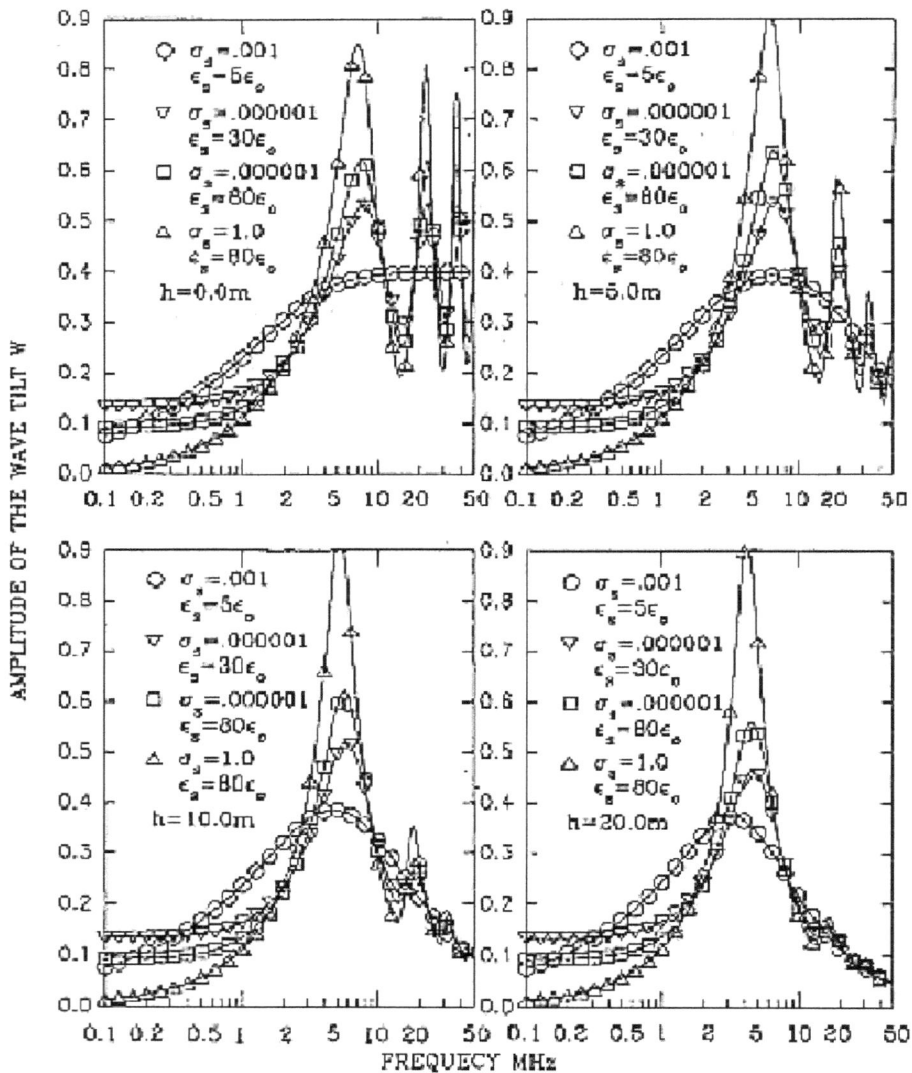

**WAVE TILE DATA
FROM WAIT AND NABLUSI
FIGURE 5-6**

For lossy dielectrics such as dirt, the incoming ray bends more toward the normal to the surface, causing an increase in the effective angle between the ray and the ground. The effect was called *wave tilt* by Wait and Nablusi. [6] They published values of wave tilt: See Figure 5-6.

Their interpretation of tilt is that a vertically polarized electric field will produce a horizontal field equal to the tilt times the vertical field intensity. Notice that the tilt is over 0.2 for the frequency range of 5 to 20 MHz, and is frequency dependent. Thus, an incoming wave glancing the surface will produce a horizontal field 0.2 times as strong or more, at HF. Simple trigonometry, or a bit of sketching on paper, shows that the effective angle of arrival is 11 degrees. A wave arriving at an angle of 6 degrees will have a total effective angle of 17 degrees.

Figure 5.7 is a calculated pattern for a quarter-wave vertical with four radials on 20 meters. Although its pattern shows a gain of -6 dBi at a takeoff angle of 6 degrees, the wave tilt converts the arriving ray's angle to 17 degrees, and the antenna gain is closer to -0.5 dBi, a gain of 5.5 dB. Interesting.

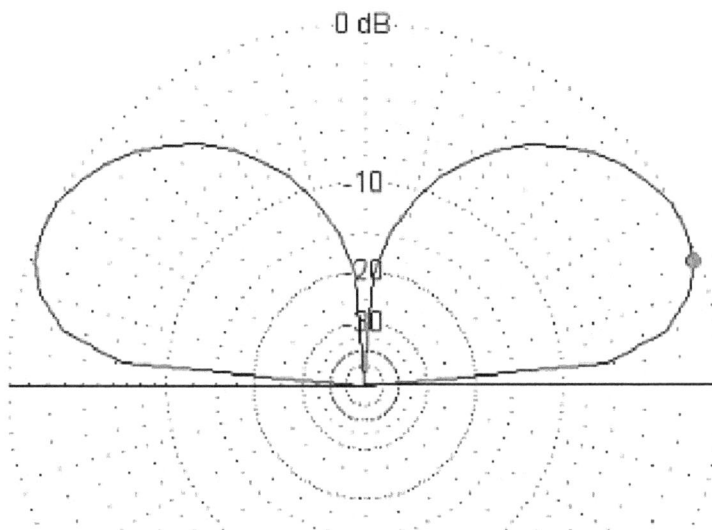

2 dB steps 6 dBi max
FOUR – RADIAL GROUND PLANE
RADIALS 1' ABOVE GROUND
e = 5   s = 13 mS/m
GAIN AT 6 deg TOA   −6.0 dBi
GAIN AT 20 deg TOA = −0.51 dBi
FIGURE 5-7

---

[6] Geophysics, Vol, 61, No, 6, Nov-Dec 1996, pp 1647-1652.

NOTES

SIMPLE BEVERAGES

Unrelated to beer and soda cans, and named after their early proponent, beverage-type antennas are a class of very low, long-wire antennas. *Low* is a relative adjective, and in this usage refers to antennas that are much less than a quarter wavelength off the ground: typical heights range from about ten feet down to ground level. These antennas are often considered to be useless for transmission, but are generally used for HF reception with great success. At the end of this chapter and in the next I discuss use of beverages for transmission as well.

I've mentioned that antennas are polarized. They can be polarized horizontally, vertically, or at some other combination of these two. We determine the antenna's polarization by looking back at it from the receiver - or, in the case of HF antennas, from the ionosphere. If the wire appears to be parallel to the horizon, it is horizontally polarized. If the wire appears to be perpendicular to the horizon, it is vertically polarized. So here's a simple question: what is the polarization of a 20 meter dipole?

The answer depends on your viewpoint. If you look at it "broadside," you see a horizontal wire: it is horizontally polarized. If, however, you look down on it from the ionosphere so you look at it almost "end on," you see a bit of *vertical* wire: radiation in this direction is vertically polarized! An interesting, but true, concept - horizontal dipole antennas radiate vertically polarized signals toward the sky off their ends!

Beverages, too, radiate horizontally polarized fields when seen broadside, and vertically polarized when viewed from the ionosphere "end on." The strong horizontally polarized reflected ray effectively cancels radiation broadside to the antenna. Off the end, however, the weak vertically-polarized reflection coefficient, at low takeoff angles, makes these antennas directional.

Electrically, beverages appear to be mounted in outer space! As a result, they do not radiate broadside to the wire, but fire *off the ends* and can be effective for DXing at HF frequencies.

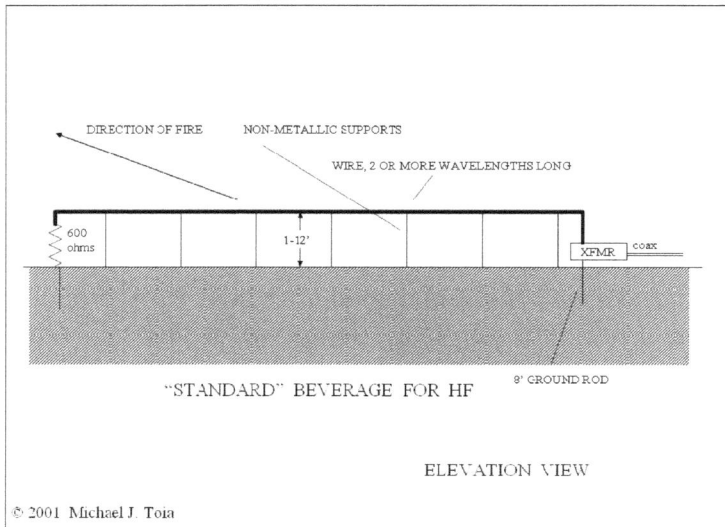

BEVERAGE ANTENNA
FIGURE 6-1

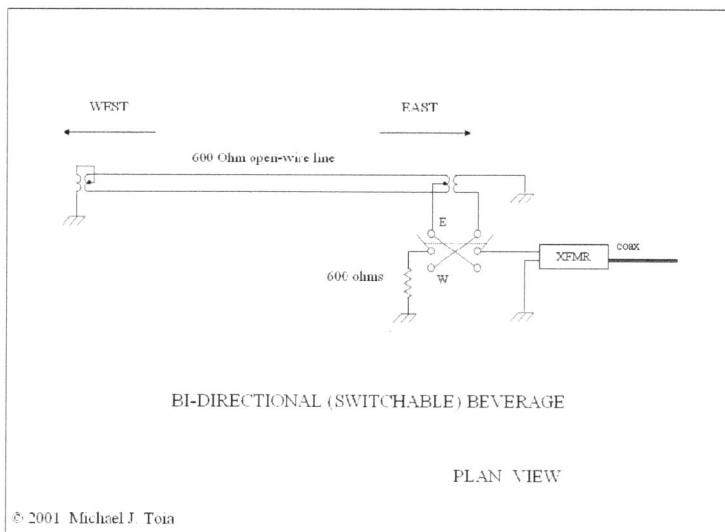

BIDIRECTIONAL BEVERAGE
FIGURE 6-2

There are several wrinkles to installing and using beverages. In general, they are made several wavelengths long, fed power at one end, and are terminated in a resistance of 600 ohms at the other end. The wire itself has a characteristic impedance against the ground, and behaves somewhat as a transmission line of 600 ohms impedance.

Three ways of installing beverages are shown. Figure 6-1 is the "typical" beverage. A wire, at least a few wavelengths long, is mounted on insulated posts a short distance above ground. Typically the height is just enough that you can walk under it to mow grass or get about the property. 7 to 10 feet is common. One end of the wire is fed with coax and a 50 ohm to 600 ohm transformer, which can be a trifilar (50:450 ohm) or quadrifilar (50:800 ohm) "balun" on a powdered iron or ferrite core. Information on these transformers is found in various handbooks and magazine articles.

The far end of the wire is terminated in a 600 ohm resistor. This causes the antenna to be broadband, and absorbs energy induced on the wire arriving from the fed end: it makes the antenna unidirectional, with a main lobe off the *far* end, as shown. The resistor can burn up when used for transmitting, unless it is designed to take the full output of your transmitter.

A second beverage *which I have not tried* but whose design appears in a few books and magazine articles is shown in figure 6-2 in a plan view (looking down from above). This antenna is similar to figure 6-1, but is made of a pair of wires spaced apart about 6" or so, to form a 600 ohm open-wire transmission line. Each end of the line terminates in a 1:1 RF transformer whose primary windings are center tapped as shown. A reversing switch allows the operator to choose a main lobe to the West (far side of the antenna) or to the East (near side) - assuming the wire runs from the East to the West, the coax on the East side. Of course, you can put this wire along any azimuth desired, to make a lobe, say, to Europe or Africa, or your choice of directions.

The action of this wire is as follows. An Eastbound wave "pushes" electrons equally in both wires to the East: it drives the transmission line in *common* mode. This induces opposing currents in the East transformer, so no voltage is coupled to its secondary.

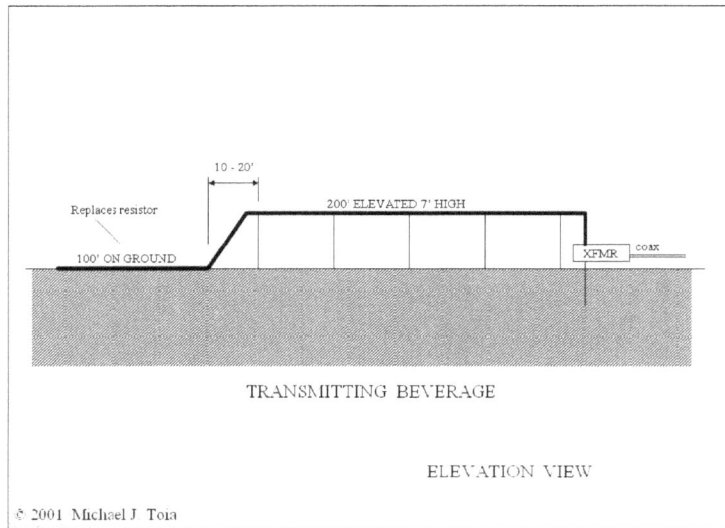

SELF-TEMINATED BEVERAGE
FIGURE 6-3

A voltage appears at the center tap of the primary, though. When the switch is "UP," in the "E" position, the voltage is terminated to the load resistor: when it is "DOWN," in the "W" position, the voltage is applied to the receiver transformer. In both cases the line is terminated in a 600 ohm load, so no energy reflects to travel back to the West transformer. With the switch in the "DOWN," or "W," position, the antenna listens to Eastbound signals - those arriving from the West. Westbound waves do exactly the same thing, but push electrons in common mode to the West transformer, where they emerge from the center tap and flow through its secondary winding. This drives the transmission line in *balanced* mode (as a transmission is usually used). As a result, this voltage is coupled to the primary of the East transformer, and develops a voltage across its secondary.

With the switch "DOWN," this voltage is terminated properly, and no reflected energy travels back to the West end of the antenna. With the switch "UP," the voltage enters the receiver coupling transformer. Hence the antenna responds to Westbound signals - those arriving from the East - when the switch is "UP" but not when it is "DOWN."

I've experimented with beverages and tried to use them for transmitting as well. The first problem encountered is the need of a terminating resistor that can handle considerable RF power.

47

Various means can be used to provide this resistance, but it's important that it be non-reactive. Light bulbs are inductive, and change resistance considerably with their temperature. Arrays of half-watt carbon resistors can be expensive. By and by I hit on a rather novel approach to the resistance, as shown in figure 6-3.

In this design the resistor is eliminated, and the termination is provided by 100 feet of extra wire laid on the ground. As I will describe in the next chapter, wires on the ground are largely resistive, with little reactance, and of about the proper value to terminate a beverage antenna. The ground wire is, of course, insulated: I commonly use #12 THHN electrical wire available at building supply centers or electrical supply outlets.

The original antenna of this type was installed along a line of wood 4x4 fence posts running E-W along my property line. The wire was my favorite #12 THHN electrical wire. A 9:1 "XFMR" wound on a powdered iron RF toroid was used to match the antenna to 50 ohm coax. Using a barefoot Drake TR-7 at about 100 W output, I was surprised to make contacts in Ohio through Missouri on 160 meters - from my QTH near Washington, DC. The antenna did what I expected - it produced a lobe to the West at a reasonable takeoff angle. In addition, when the 1BCG commemorative station was on the air a few years back, I tried each evening to work him - and succeeded on the seventh and final evening of that test! The commemorative station was reportedly near the original 1BCG site at Greenwich, CT. This antenna was NOT oriented properly for that contact, but did get through a crack in the din of all others that were calling.

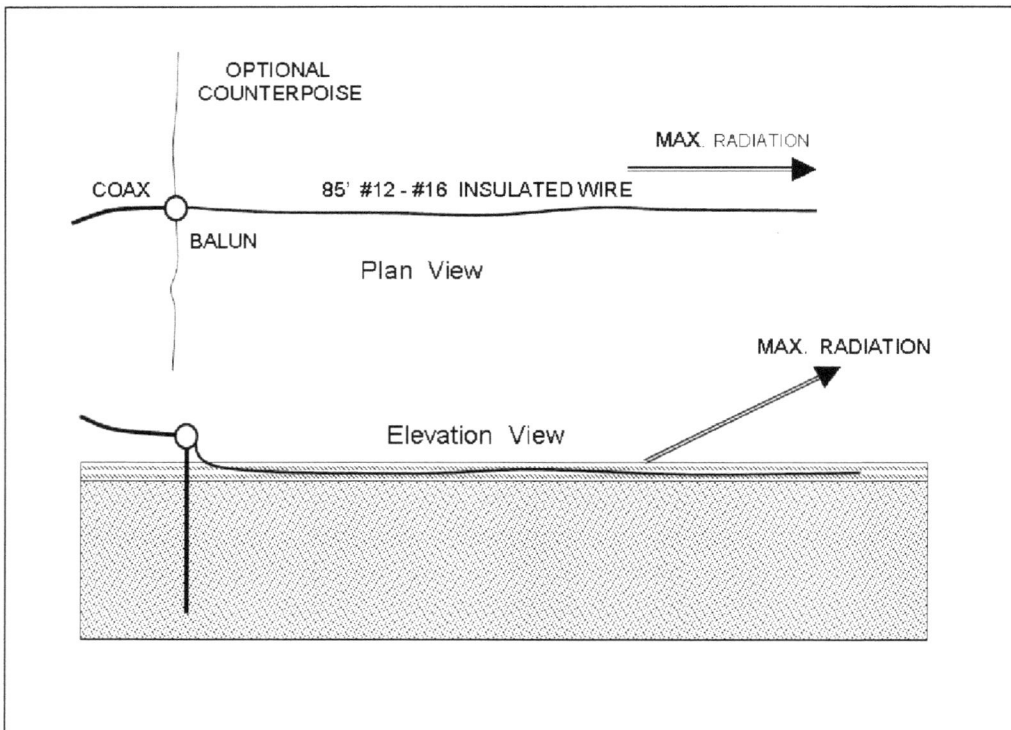

ELEVATION AND PLAN VIEWS OF THE "GRASSWIRE"
(A VERY LOW BEVERAGE ANTENNA)
FIGURE 7-1

7

THE "GRASSWIRE" [7]

LOW BEVERAGE ANTENNA

Deed restrictions got you down? Neighbors intimidating your tower plans? Need a really easy, portable HF antenna? Then the Grasswire may be the answer! Virtually invisible, lightweight, and compact (you can carry one in your hip pocket), this antenna works! I've used one in various installations for more than 10 years.

Read on - and listen to the "experts" telling you that this is hogwash, that an antenna like this can't work. But it does. And true experts, who have taken a decade or more to come to grips with the intricacies of Maxwell's Math, know why.

Grasswires will not out-perform a yagi, or a decent dipole up a half wavelength. Not in gain or signal strength, at least. But they do survive ice and wind storms, and are practically immune to lightning. And they don't need a large tower or tall support. I deploy one from my hip pocket at times - the balun to match it is larger than the antenna!

The Grasswire - In Brief

What is it? Put simply, it's an end-fed, longwire antenna that is laid right on the grass. Hence the name. Look familiar? Yup! It's a very low beverage antenna, as shown in figure 6-3 of chapter 6.

My first Grasswire, built in the summer of 1988 was just 204' of #22 AWG magnet wire laid along the property line, anywhere from 1" to 6" above the ground. Figure 7-1 shows plan and elevation views of a typical installation.

----

[7] Recent reports of this type of antenna, unaware of my earlier papers on the matter, have dubbed this antenna the *BOG*, for *Beverage on Ground.*

An 8' ground rod and optional counterpoise wires are shown. The counterpoise is a 40' wire, center tapped. Use it or the ground rod: you don't need both.

These antennas are largely resistive, with values ranging from 150 to 500 ohms or so on average ground. I've used them successfully on the soils northwest of Washington, DC, on the sandy soils of the Cape Canaveral, Florida area, in the rocky, shale soils of the mountains in Somerset county, PA, and on river bottomland of Allegheny County, PA. One was used with great success by K3MT/VP9 in Southampton, Bermuda - the object of nightly pileups on 30 m CW for four nights.

Reflection And The Brewster Angle

The skeptic in you will doubt that such low antennas can work. After all, its image in the ground radiates and cancels out all radiation. True - if the ground is perfect. But nothing is perfect! The Grasswire is a low beverage antenna (see chapter 6), and as such, radiates vertically polarized off its end. Extensive monitoring tests with wires laid along the great circle route toward WWV and perpendicular to that line demonstrated the end-fire nature of the antenna. If you've read and understand chapters 5 and 6, you understand why this antenna works. Here's a recap:

When a radio wave reflects from an air-earth boundary, an incoming ray reflects, giving an outgoing ray. These two, and the line normal to the boundary plane, form a plane of incidence. Solutions of Maxwell's equations differ for the case of the E-field being perpendicular to this plane (i.e., horizontally polarized), and the case when the E-field vector is in the plane of incidence. You will probably call the latter "vertical" polarization, although this is technically not correct. Electromagneticists (a.k.a. those who practice *Electromagical effects* [8]) refer to these cases as perpendicular incidence (horizontal polarization) and parallel incidence (vertical polarization).

As I've discussed in chapter 5, horizontal polarization reflection is nearly total, with a nearly 180 degree phase reversal. So very low antennas neither respond to, nor generate, appreciable amounts of horizontally polarized radiation. But for "vertical" polarization, the reflection varies in strength considerably. At some takeoff angle (angle between outgoing ray and the ground) the reflection becomes quite weak, and has a 90 degree phase shift.

---

[8] A bit of shop humor

Near this angle, the sum of direct and reflected rays will have a magnitude as if the antenna were in free space! Of course, at other angles, ground reflection largely cancels the direct ray, and the antenna does not radiate well at all.

Look back at figures 5-4 and 5-5 of chapter 5, at the curves for vertical polarization. Notice that, at 10 to 25 degrees, the ground reflection is very weak. It is also shifted nearly 90 degrees in phase from the incident ray. Therefore, radiation from the Grasswire, off the ends will be about the same as if the ground were not present - as was discussed earlier for beverage antennas.

But launching a ray at, say, 15 to 20 degrees takeoff angle, in a direction toward Europe, can be useful! The concept of wave tilt can reduce the takeoff angle buy several degrees. That's what a Grasswire does. It is lossy in all directions, but least lossy when exciting the ionosphere for a long-haul DX contact.   To demonstrate the point, here's an extract of my log for October of 1988, (ahh, glory! Yes, the sunspot number was good then!) using a Grasswire:

Date	GMT	CALL	his/my RST		FREQ	Power	
OCTOBER							
27	1554	SM6DYK	579 /	559	28004	80	
	1601	SM0LBR	569 /	439	21007	100	RAY - STOCKHOLM
	2001	W4JBQ	579 /	569	7029	40	JOE - FT WRIGHT, KY
	2141	W8LNJ	579 /	459	28015	80	DAVE - DALLAS, TX
28	0227	W8AO	589 /	569	3547	15	BOB - SILVER LAKE, OH
	1720	G3RFE	579 /	559	21016	100	TOM - BARROW
	1932	G0CBW	569 /	559	14029	50	MEL
	1945	VE2FOU	589 /	559	7032	100	ANDRE - IBERVILLE
	2026	KB7UX	569 /	539	21040	100	RUSS - AZ
	2100	I2JIN	589 /	559	14022	40	BOB - COMO
	2123	G3JVC	569 /	559	14022	40	JOHN - LONDON
29	2105	WA2OOJXT	599 /	599	28015	80	ND

Not bad, for a wire on the ground. Notice that contacts were made on 80, 40, 20, 15, and 10 meters. The signal reports are not fantastic. But contacts were made, and ham radio was enjoyed! Five countries were worked in 3 days. And the best part of this setup: the neighbors never knew that a ham station was on the air!

During our Bermuda excursion, I took a TR-7, small antenna tuner, a power supply, and a Grasswire. We were guests of a small family-run group of rental cottages in Southampton for four days. On the third day, one of the elder family members chatted with me a bit, and asked if I was perhaps a radio amateur.

Construction

A1

B1

C1

A2

B2 C2

T-200-2 Core

Wiring

C2
C1
B2
B1
A2
A1

Coax "hot"	Grasswire	Antenna Z
A2/B1	B2/C1	200 Ω
A2/B1	C2	450 Ω
B2/C1	C2	125 Ω

Trifilar Balun:
use #16 - #22
insulated wire

K3MT
(c) 1997
M. Toia

© 2001 Michael J. Toia

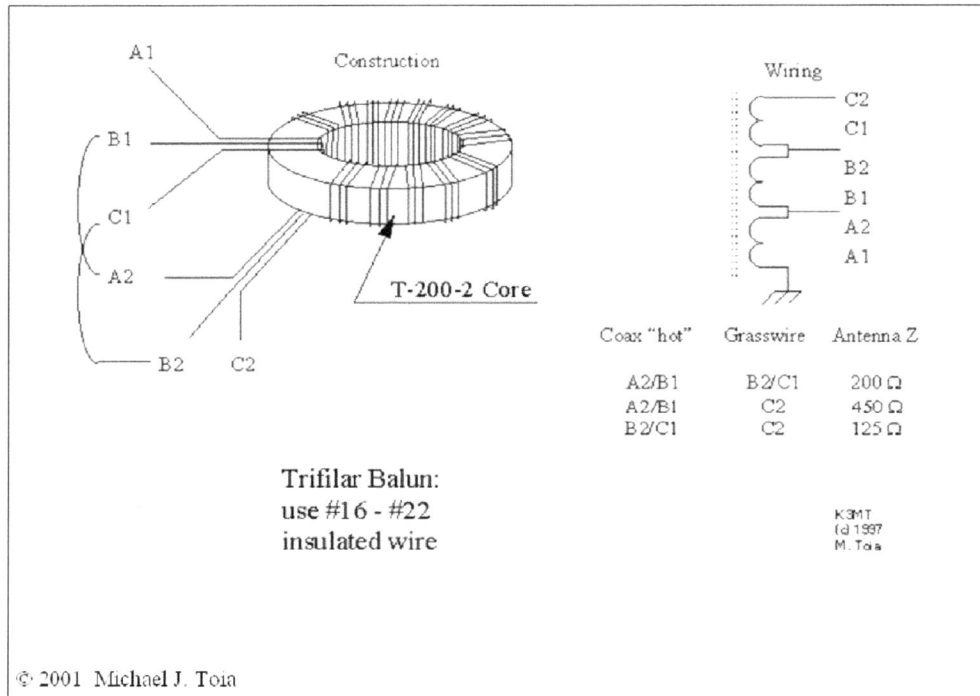

A TRIFILAR (3-WINDING) BALUN FOR FEEDING THE GRASSWIRE
WINDINGS ARE 16 TURNS OF THREE PARALLEL WIRES
FIGURE 7-2

I said, "yes." In fact, I had been on the air for the previous two days, using a Grasswire. It was only that morning that he, in cleaning up around the area, came across the antenna! *That's* a low-profile antenna.

Feeding the Grasswire

Since this antenna is largely resistive, a simple trifilar balun is all that I have ever had to use. Figure 7-2 shows how to make a balun that works.

Typically I pull the insulated jacket off some indoor telephone wiring cable. Four insulated #22 copper wires are inside: discard one of these and use the remaining three. Wind about 16 turns on a T-200-2 core (available from Amidon and others), without allowing the wire to twist (keep the three conductors parallel at all times.)

Notice that this "balun" really matches an unbalanced antenna to an unbalanced transmission line. It is basically a wide-band, three-winding autotransformer. Impedance ratios are as shown on the drawing. Generally it is necessary to connect the coax to either A2/B1 or B2/C1, and the antenna to B2/C1 or to C2. This may change from one band to another, and usually does.

How much wire?

A general rule of random wire antennas is to get as much wire in the air as you can - longer is better. Does this still hold for the Grasswire? The answer is no. Measurements show that anything over a wavelength does no appreciable good.

My first measurement program parked a car on a dirt trail, with a spool of 18 gauge insulated wire unwound, one end tied to the bumper and the rest run on down the trail. The dirt was average stuff, mostly clay and loam on top of granite. At the car the wire was untied from the bumper, passed through a small RF toroid, and connected to an antenna tuner, the latter driven by a TR7 transceiver at approximately fifty watts. The car itself served as a counterpoise.

A ten-turn secondary winding on the toroid drove a small diode and capacitor. RF current in the antenna developed a DC voltage across the capacitor that I measured with a handheld DC voltmeter. As the toroid slid along the wire, the voltage dropped and fell below 10 percent of the starting value a wavelength along the wire.

54

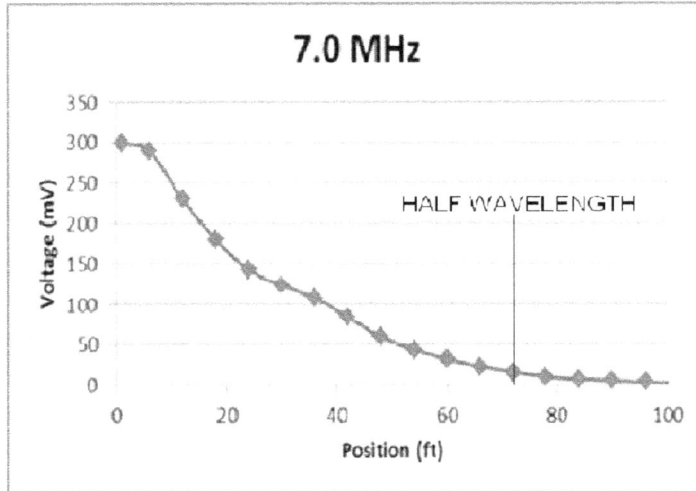

CURRENT PROFILE AT 7 MHz

FIGURE 7-3

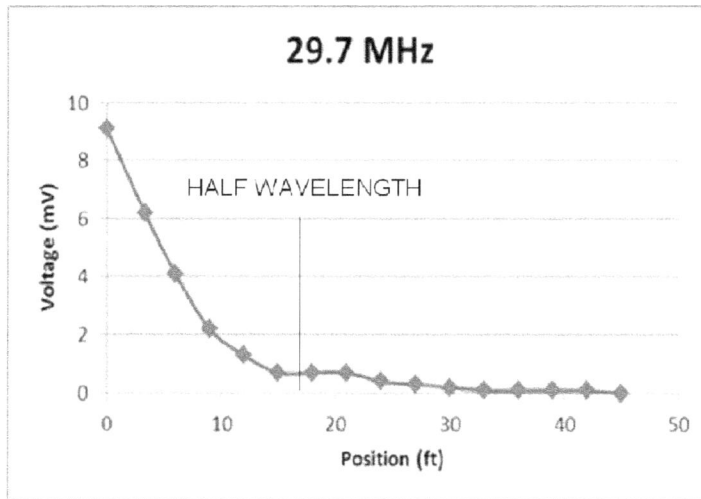

CURRENT PROFILE AT ~30 MHz
FIGURE 7-4

There was a small rise in voltage for a short bit farther along the wire, but at a full wavelength it fell below one percent, and never showed any further improvement. This occurred on 80 40, 20, 15, and 10 meters.

This measurement indicated that the current in the wire dropped almost exponentially along the wire, and beyond a wavelength was more than 20 dB down, so could produce little radiation. The excess wire can simply be removed. Thereafter my Grasswire deployments always used about one wavelength of wire at the lowest operating frequency.

Continuing the measurement at a later date, an assistant and I laid a center-fed wire dipole on a grassy field, 396 feet of insulated, 12 gauge wire - all that we happened to have handy. Again a small toroid RF transformer and diode/capacitor, similar to the earlier one, had one side of the dipole threaded through it. A fiberglass surveyor's tape stretched from the center along the dipole to one of its ends. The DC voltage, measured as a function of distance along the wire is a measure of the RF current. Figures 7-3 and 7-4 show the falloff of current along the wire, and its attenuation by at least 20 dB at one wavelength (seventy, and sixteen and a half feet, respectively.)

Windom In The Grass

In chapter 3 I've described a windom antenna. While it is usually hung from a pole or in a tree, it works when used in a "Grasswire" mode. Just lay it on the ground. Dimensions are repeated in figure 7-5 for ready reference.

When I travel, I sometimes take one of these made of #22 insulated hookup wire. Since I often set up beside motel parking lots, and often after a day's work, sometimes after dark, I've color-coded the wires. The longer is black, and the shorter, red. This helps me determine which way to point the windom. Remember, though, that it fires off the long end. Of course, it fires the other way, too, but with a bit less gain.

I hope this has given you a good case of curiosity. Go out and try one of these ground - mounted wires. They're easy to build. Even the balun is easy to build. And they're fun to play with.

A last word to put off discouragement: the Grasswire, and beverages in general, are lossy compared to a good dipole at a half-wave height. They are disappointing for QRP operation. Run a bit of power with these antennas. 100W RF or more should do the trick.

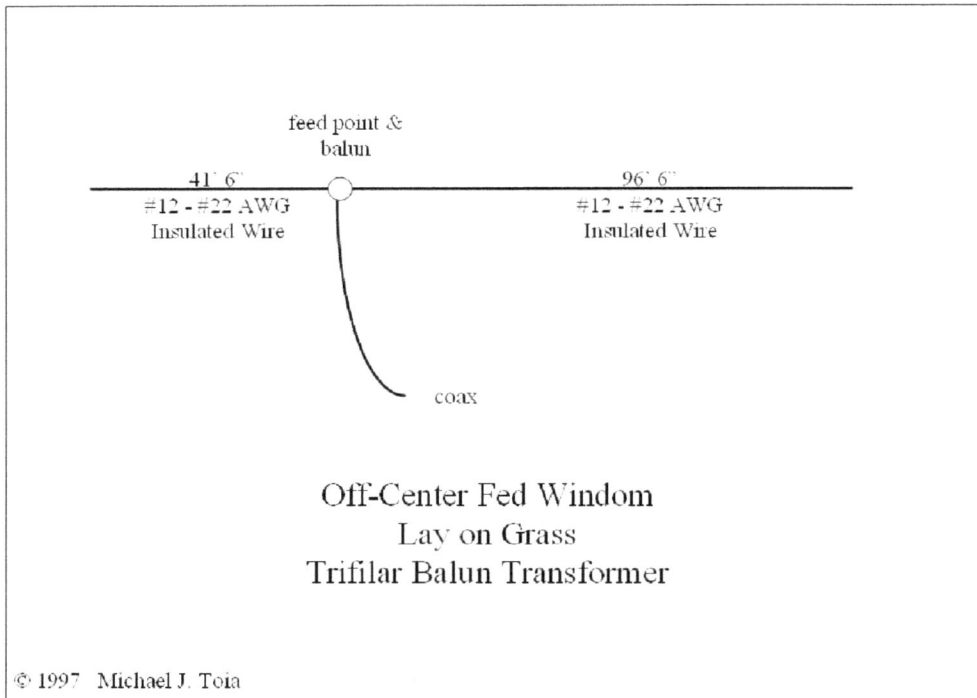

feed point &
balun

41' 6"
#12 - #22 AWG
Insulated Wire

96' 6"
#12 - #22 AWG
Insulated Wire

coax

Off-Center Fed Windom
Lay on Grass
Trifilar Balun Transformer

© 1997 Michael J. Toia

A WINDOM (CHAPTER 3) ON THE GROUND
FIGURE 7-5

8

VERTICALS THAT WORK

AND WORK WELL

Verticals tantalize. Antenna handbooks say they have maximum radiation along the earth, at a very low takeoff angle. They ought to be the best simple antenna for working DX.

Working DX is like firing a cannon. Fire at a very high takeoff angle, and hit yourself. Fire at a lower takeoff angle to hit a more distant target - the lower, the farther the cannonball goes. [9] Ardent Dxers want as low a takeoff angle as possible: they want the signal to bounce off the ionosphere and come down somewhere in that elusive foreign country. That's the vertical's appeal.

Alas, despite the simple handbooks, verticals *do not* radiate at zero degrees takeoff angle (right along the horizon): this is a stunning and popular misconception. As I've shown in chapter 5, the ray reflected from the ground is very weak at a particular angle, about 5 to 15 degrees at HF, depending on frequency and soil parameters. Below this *Brewster angle*, the reflection becomes 180° out of phase and very strong for verticals, just as it does for horizontal antennas. This reflection cancels radiation at very low takeoff angles for ground-mounted verticals. But verticals *can* do a good job of DXing if installed properly.[10]

---

[9] Using the "copper ionosphere" model. To do this, you must change your operating frequency with increasing takeoff angle, so the "cannonball: will always reflect from the ionosphere.

[10] OK, let's clear the air for you deeply-oriented electromagneticists. What I've said states that ordinary AM broadcast transmission cannot work. The hitch is this: if ONE of the antennas, either transmitter or receiver, is high above the ground, then the reflection coefficient approximation is correct. When BOTH are within a wavelength or so of the ground, a modification of the theory shows the existence of a "Ground Wave" that carries energy from one antenna to the other. For amateur work on HF, I consider one antenna to be the ionosphere - quite high above the ground - out of range of the ground wave.

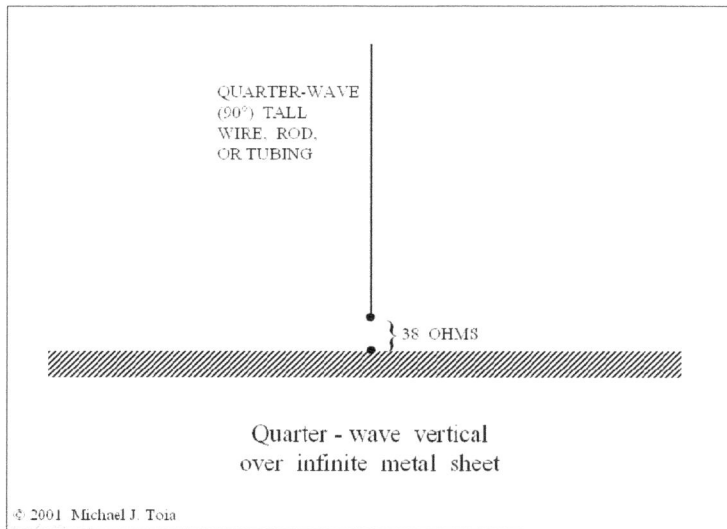

**IMPEDANCE OF A QUARTER-WAVE VERTICAL
OVER PERFECTLY-CONDUCTING GROUND
FIGURE 8-1**

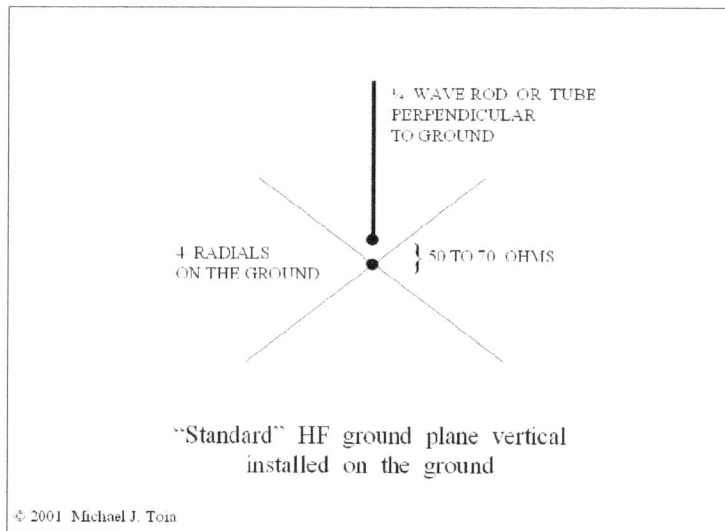

**IMPEDANCE OF A QUARTER-WAVE VERTICAL IN THE BACK YARD
FIGURE 8-2**

The input impedance of a classic quarter-wave vertical antenna mounted above an infinite, perfectly conducting metal sheet is 38 ohms resistive, with zero reactance. See figure 8-1. If your property is paved with copper, silver, or aluminum for several wavelengths in all directions, a vertical will work well. But most hams can't afford that much copper. Broadcast stations *can*: they bury a minimum of 120 wires in the ground running radially outward from their AM towers for about a quarter wavelength.

The radio ham, on the other hand, tries to use just four such radials. The result is an input impedance of 50, 60, or 70 ohms or so, depending on soil conditions. See figure 8-2.

Suppose the impedance is 60 ohms. 38 ohms of this is the antenna's resistance (that produces the radiated signal) while 22 ohms is the ground resistance. With a transmitter power output of 60 watts, a current of one ampere will flow into the antenna. ($P = I^2 R$) The ampere into 38 ohms actually radiates from the antenna, with an effective power of 38 watts. The ampere into the remaining 22 ohms, or 22 watts, will disappear into the ground system. A wee bit of this actually radiates, but most of this power just heats the worms. The efficiency of this antenna, the radiated power divided by total applied power, is 38/60 = 63%.

The vertical antenna looks the same from every point along the horizon, so it radiates in all directions. I.e., it is *omnidirectional.* But its ground losses rob some power from the transmitter, and ground reflections contribute nothing to the signal below the Brewster angle. By contrast, the horizontal dipole installed a half wavelength high or more has its radiation **reinforced** by ground reflection, giving it an inherent doubling of E-field strength, or a **6 dB increase** [11] over a lossless vertical. The vertical gets the reputation of radiating "uniformly poorly in all directions!"

How can this simple antenna be made more effective? There are two basic approaches. One is to do as the broadcasters do, and plant a lot more radials. This lowers the ground impedance and improves efficiency, **but** - the height of the currents on the antenna does not change. The main effect is to recover power lost into the ground - a dB or two. Since many other papers talk about improving HF verticals by adding lots of radials, I drop the issue here. In fact, I'll present a means of *reducing* the radial count and yet *improving* performance of the vertical.

---

[11] For perfectly conducting ground. Typical values are closer to 5 dB - still a good advantage over verticals!

The trick is to study the experimental measurements of *Brown and Woodward*, reprinted in Jasik's *ANTENNA ENGINEERING HANDBOOK, first edition, 1961,* by McGraw-Hill, specifically its Figures 3-3 and 3-4. These show measured values of input resistance and reactance for cylindrical verticals mounted over a very large ground plane, for cylinders of differing *length to diameter* ratios (L/D). The values for L/D = 1000 are taken as typical for the small-diameter verticals we radio amateurs use.

To describe these measurements, I need to introduce a new way to measure length. We usually think in terms of length in feet or meters for HF work. But I will use the unit of *degrees* of length. Most broadcast engineers, and Brown and Woodward, describe antenna dimensions in degrees. A degree of length is $1/360_{th}$ of a wavelength, or one wavelength = 360°. A half wave dipole is 180° long, a quarter-wave vertical is 90° high, and so on. For example, at 3.75 MHz, the wavelength is 80 meters, and a quarter wave(length) vertical is 20 meters, or 90°, tall.

A thin vertical about 85° tall mounted above a large, conducting sheet, is at resonance according to the measurements: its reactance is zero. Its input impedance is purely resistive, and about 38 ohms. As it gets taller than 90°, its input resistance rises, and its reactance becomes *inductive*. At about 175°, its reactance suddenly shifts from inductance, through zero, to capacitance, and its resistance is over a thousand ohms. If we were to make a vertical nearly 175 degree tall, tuned to an input resistance of 978 ohms, and operate it against a ground of four radials with a ground resistance of 22 ohms, the total feed point resistance would be 1000 ohms. A 10 watt transmitter would produce 100 mA in this resistance. 100 mA into the 978 ohms radiation resistance would produce 9.78 watts of radiated power, while the 22 ohms of ground resistance would dissipate only 0.22 watts! The efficiency of this vertical is quite high - 97.8% of applied power goes into radiation.

There's a bonus and a pitfall of the 175 degree vertical. First the bonus - the maximum current is now half way up the antenna, well off the ground. This works with the *reflection coefficient* to lower the takeoff angle a bit. Now the pitfall - the high *reactance* must be tuned out with a series coil or capacitor, which has some additional loss. Furthermore, the rapid swing in reactance from largely inductive to largely capacitive makes this vertical unstable in tuning. In fact, across say, the 80-75 meter band the antenna may swing from being inductive at 3.5 MHz to capacitive at 4.0 MHz. (Yes, that's right - inductive at the lower frequency.) But we are *definitely* on the right track!

Increasing the height of the vertical to either of two values - 136° or 3/2 of that, 204° - yields an input impedance of about 200 to 300 ohms resistance with either inductive (136°) or capacitive (204°) reactance. The reactance stays fairly stable with small frequency changes or changes of weather.

A 204° (tall) vertical will have its current pattern higher off the ground than the shorter 136° vertical, but needs a taller support. It requires a series loading coil to tune out its reactance. The 136° vertical, on the other hand, needs only a series *capacitor* to tune out its reactance. Let's examine these two tuning methods and see how they affect efficiency.

According to Brown and Woodward's measurements, the 136° vertical has Z = 300 + j 450 ohms - 300 ohms resistance in series with 450 ohms inductive reactance. This will require a series capacitor of 450 ohms reactance. Capacitors have a Q (quality factor) of about 1000, so the capacitor will present a series resistance of 1/1000th of its reactance - about 0.45 ohms. This dissipates only a tiny fraction of the total applied power, so efficiency does not change enough to be measured by even sensitive techniques.

The 204° vertical, on the other hand, has Z = 230 -j 550 ohms (it is *capacitive* by 550 ohms). It requires a series *coil* with reactance of 550 ohms. Good coils have Q's on the order of 200 or better, so the series resistance due to the coil is 550/200 ohms, a bit less than 3 ohms. This is slightly higher than the series capacitor's loss resistance, but not significant, because it is only about 1.5% of the radiation resistance. Again it is a bit difficult to measure the loss of efficiency due to this coil. Therefore, either the 136° or 204° vertical would be a good antenna to try.

I routinely use both these antennas on 10, 12, 15, 17, and 20 meters with one travel pack, consisting of a home-brew tuner, a Yaesu FT-747 HF transceiver, and other stuff packed into a small travel case that fits under the seat in front of me on airplanes. So what is my ground system?

I use a *single* quarter-wave wire laid on the ground as a ground lead. It has an impedance of some 90 ohms, more or less. Compared with the 230 ohms radiation resistance of the 204° vertical wire, it causes a small loss of efficiency. For this antenna, the total input resistance is 320 ohms. A 320 watt transmitter would produce one ampere into this load. $230/320_{ths}$ of this power (230 watts) goes into radiation, and 90 watts goes into ground loss. The efficiency is 230/320 = 72%. For the 136° wire, the total feed resistance is 390 ohms.

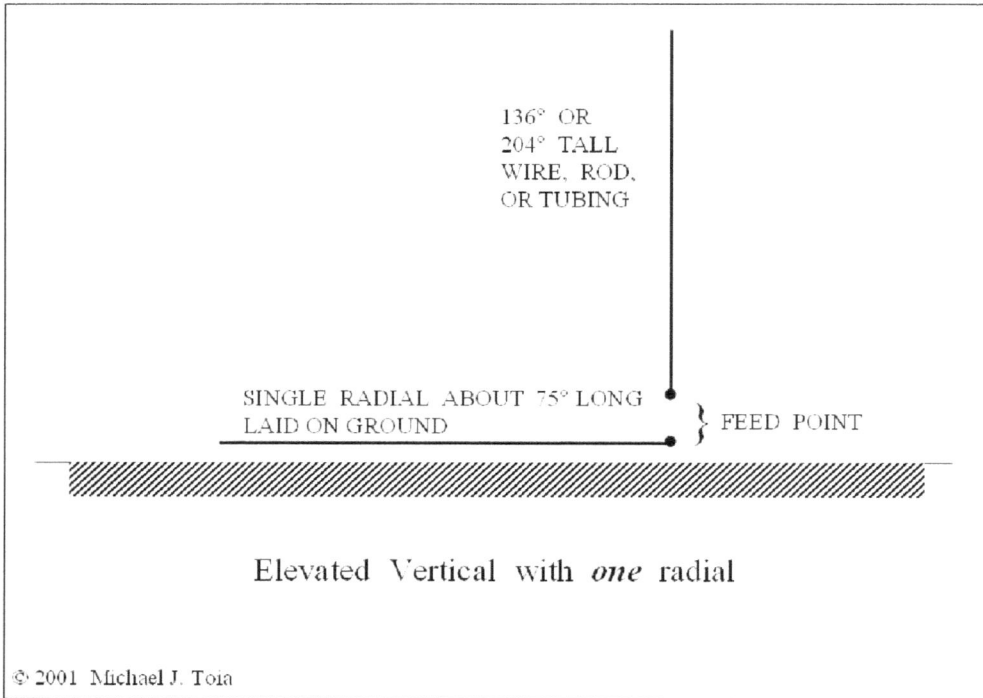

136° OR 204° TALL WIRE, ROD, OR TUBING

SINGLE RADIAL ABOUT 75° LONG LAID ON GROUND

} FEED POINT

Elevated Vertical with *one* radial

© 2001 Michael J. Toia

**A SIMPLE, EFFICIENT VERTICAL USING A SINGLE "RADIAL" WIRE.
FIGURE 8-3**

A 390 watt transmitter would produce 300 watts of radiated power and 90 watts of ground loss. While this is a bit more efficient (77%) , the point of maximum current is closer to the ground, leading to a slightly higher takeoff angle. Therefore I deploy the 204° configuration when I can, and fall back to the 136° wire when I cannot.

A note on the ground wire: wires laid on the ground resonate at lower frequencies than those up in the air, due to the large capacity between the wire and ground. Thus the "quarter wave" radial is really about 75° long: its dimension is not too critical, and anything from about 70° to 140° works well.

Figure 8-3 shows the typical "minimal" installation of the 136°/204° vertical. By placing three more radials on the ground, the ground impedance can be reduced to the order of 20 ohms. With a *single* radial, the efficiency of the 204° wire is 72%. A loss of 1.4 dB is incurred with this simple ground, vice a loss of 0.36 dB for a 4-radial ground. Paving the county with copper would recover about 0.35 dB of this loss - not worth the effort!

These elevated verticals are interesting, easy to build, and more efficient than the standard, quarter-wave-over-four-radials vertical. Give them a try - you'll be surprised.

NOTES

9

INVERTED U's

FOR 160, 80, AND 40

Do you find it hard to get on 160 meters because of the antenna? This simple wire may be just what you need. It installs easily, needs no elaborate ground system, yet performs very well. A little ingenuity solves two conundrums at once - getting the current high in the air, and lowering the current (and hence losses) in a very simple ground system. You'll find this antenna to be a variant of the verticals discussed in chapter 8. In fact, I developed this one, and based on its performance, went on to build my travel antenna and the designs of the previous chapter.

The Usual Problems

160 meter antennas are large. One way to meet the size problem is by wrapping a loop antenna around the house, tuned to resonance with a capacitor. With today's electromagnetic safety concerns (see the FCC rules on the subject) this may not be the best way to meet RF exposure limits.

Another approach is a vertical whip with center and/or bottom loading coils. But a resonant 160 meter vertical is more than 120 feet high. That's a tall order - literally! Shorter whips need to be tuned to resonance with coils, and these are lossy. The best coil may have a Q of 330 or so, and many home-brew coils have Q's much lower than that. It's not uncommon to need 1000 ohms of inductance from a coil of Q=200, which puts 5 ohms of resistance in series with the antenna. Adding insult to injury, the antenna's radiation resistance may well be below 5 ohms, and usually is. Therefore, most of the power goes into heating the coil - and the ground system.

To get maximum radiated power at a good takeoff angle, the antenna current must be as high, both in amperes and elevation above ground, as possible.

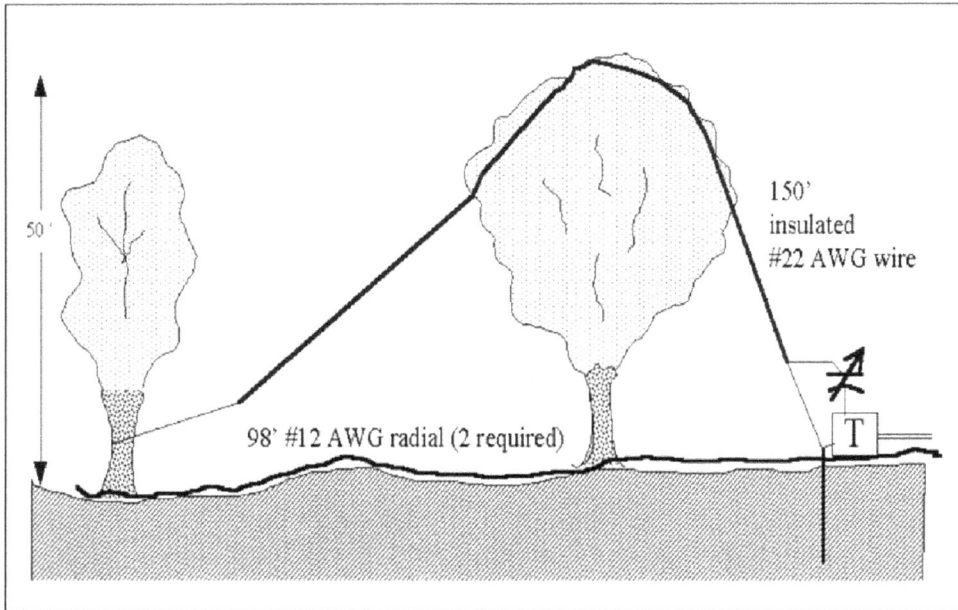

50'

150'
insulated
#22 AWG wire

98' #12 AWG radial (2 required)

T

THE 160-METER INVERTED "U" ANTENNA
FIGURE 9-1

Base-loaded verticals have a current maximum at the ground. Therefore, the ground system must 'sink' this current. Every ohm of ground resistance is critical, as the $I^2 R$ loss in the ground is lost power In standard (AM) broadcast antennas, low ground resistance is accomplished by using a minimum of 120 radials spaced at 3 degree intervals. For amateur work, this is again a formidable undertaking.

A Bit of Ingenuity

If a 160 meter whip were made more than 135 feet high, the point of maximum current would no longer be at the ground; its feed impedance would be greater than the quarter-wave 38 ohm value, and could be inductive, not capacitive. A series capacitor would be needed to bring it to resonance. This was discussed in some detail in chapter 8.

Now, capacitors have higher Q's than coils - they can easily be more than 1000. Since a "taller than quarter wave" vertical will both exhibit a higher radiation resistance (and, hence, a lower feedpoint current for a given power input), and can be tuned with a high-Q capacitor, the power lost in the tuning element will be a smaller fraction of the total power than with base-loaded shorter verticals. The current through the capacitor will be lower than that through a loading coil, and its $I^2 R$ loss will be less than the loss through a tuning coil for two reasons: less current and higher Q.

The need to 'sink' a lower current in the ground system allows a simplified ground to be used effectively. This, and the other effects just discussed, lead to a really easy-to-construct vertical for 160 meters.

The Design

Some experimentation on the ideas above led me to take 150 feet of insulated #22 hookup wire and draw it through the crown of a cooperating oak tree, as in figure 9-1. The tree is about 60 feet high. I used my fishing rod, monofilament line, a one-ounce lead 'egg' sinker, and a 'wrist-rocket' slingshot to put the line over the crown of the tree. By untying the lead weight at the far end (at ground level) and attaching nylon 'chalk line,' I was able to pull the chalk line over the tree back to the starting point.

Next I tied one end of the hookup wire to the chalk line, and fastened the other end to a stake driven into the ground. Then I pulled the wire over the tree's crown and down the other side. By pulling the chalk line I was able to bring the wire very lightly taut. I fastened the line to the trunk of another tree. The free end of the wire was kept at least 10 feet off the ground for safety and to avoid excessive detuning.

A ground rod 8' long and 5/8" in diameter - the type used for electric service entrance grounding and available at many hardware stores, electrical supply houses, and some Radio Shack stores - was driven into the ground at the feed point. I tied the feed end of the wire to this stake with about 1 foot of chalk line, a "cheap and easy" insulator.

The ground system

A half-wave dipole in free space is about 260 feet long for 160 meters. When placed on the ground, its resonant frequency drops. To bring it back to resonance, cut it to about 60% of its former length.

At the same time, its feed impedance rises from 76 ohms to about 200 ohms. Therefore, I placed a 198 foot piece of insulated #12 wire on the ground as if it were a center fed dipole, with its center at the ground stake. Both wires at the center of this "dipole" were connected to the ground stake. This arrangement yields a ground system whose impedance is on the order of 60 ohms. Since the antenna feed impedance (over perfect ground) is on the order of 150 ohms, the ground loss is only about 1/3 of the total supplied power - quite a bit less than with a short vertical.

Feeding the antenna

With the simple 2-radial ground system, the antenna's combined feed impedance is about 200 ohms. To feed it, I use a trifilar balun as shown in chapter 7, figure 7-2. It is usually connected for a 125 or 200 ohm feed: see the figure for impedance ratios for various taps on the transformer. The 'balun' is actually a wide band, unbalanced, impedance step-up transformer.

The antenna wire was connected to the balun through a series tuning capacitor, an ordinary 2-gang, 365 pF variable removed from an old broadcast receiver. It works just fine without flashing over at power levels up to 120 watts or so.

I've used the antenna with a barefoot Drake TR-7 and a Yaesu FT-747. The ground end of the balun and feed coax was connected to the ground stake/radial junction point.

Set the capacitor about midway. Then check the SWR at 1.8, 1.9, and 2 MHz. Adjust the capacitor accordingly (with the RF drive OFF! - its frame will be -hot-with RF!) until you resonate the antenna at the desired operating frequency. The antenna bandwidth will be about 50 kHz or so, and you will need to re-tune the capacitor if you want to go from the low end to high end of the band.

Insulate the capacitor by placing it in a tupperware container or on a block of wood. I keep the rain off the capacitor and 'balun' by covering them with an inverted plastic dishpan or bucket. You can be more formal and build the capacitor / balun into a nice weatherproof box if you want.

80 and 40 Meter Versions

After having such good luck with this antenna, I began to investigate cutting it down to size for other bands. That led to the developments in chapter 8. Since it's hard to get a support that keeps 136° [12] of wire vertical, I usually co-opt a tree, as done with the 160 meter version, to keep as much of the wire as vertical as I can, then let the rest arc across the tree's crown and head back to the ground. Again a simple balun and series tuning capacitor are all that are needed to match the antenna to 50 ohm coax.

By the time I scale these antennas to 30 or 20 meters, I can often avoid the "U" shaped arc and keep all, or most, of the wire vertical. Thus the verticals of chapter 8 are essentially the same antenna, with added flexibility, such as being able to use 204° of wire on the higher bands.

Final Thoughts

The 160 meter version of this antenna has been used successfully to work into Europe on summer evenings! I tried it during Field Day '97, and it worked quite well. Unfortunately, 160 meters is not a good band for contesting during Field Day. But that's not the antenna's fault.

------

[12] (Very nearly) 3/8 wavelength. See chapter 8 for a discussion of wire length in degrees.

I rebuilt the original, making it 136° of #12 THHN electrical wire. This version was used on the ARRL and CQ 160 meter contests the winter of 2000-2001 with great results. In the latter contest, I worked 45 states in one evening: still need ND, SD, KS, HI, and AK. Oh, well - sometime soon, I hope.

10

MIKE VILLARD'S

NOISE - CANCELING ANTENNA

Here's a neat little antenna for receiving on shortwave - that is, HF - frequencies. It's the brainchild of Mike (a.k.a. Dr. O. G.) Villard, Professor Emeritus of Stanford, founding father of SRI Inc, and one of the most wonderful colleagues with which it has been my sheer pleasure to be associated.

A Bit of Background

Mike had been asked by the Chief of U.S. Information Agency, the parent of Voice of America, if there might be some way to counter Soviet jamming of VOA broadcasts. His mind set to work: he made a *pile* of different antennas that seemed to show some promise, and presented a proposal to USIA to produce two different anti-jam antennas that could be effective. I was at VOA engineering. His proposal became a contract: by the best of luck, engineering thrust *me* into the role of "ARCO" - authorized rep of the contracting officer (most other government agencies call this "COTR," but USIA is, well, different!)

I visited Mike at SRI frequently. He showed me myriad designs that filled his pool house to the max! We selected two promising ones, and spent the rest of the year improving them, producing two really neat end products. One counters sky wave jamming, where the jammer propagates by ordinary HF propagation. The other counters "ground wave" jammers - noise transmitters near the receiver, usually on the outskirts of the city being jammed.

Ground wave jamming is very similar to arcing power lines or other locally produced noise from appliances and the like - even from lightning. Since we hams are sometimes plagued with local noise, Mike's little ground-wave antijam antenna can be quite useful. And, it's simple to build. So, without further delay, I present, in picture form, …

The HLA - Villard's Noise Rejecting Horizontal Loop Antenna

Get a 2' x 2' base - plywood, cardboard, stiff foam plastic, even the back of a wall-hung picture.

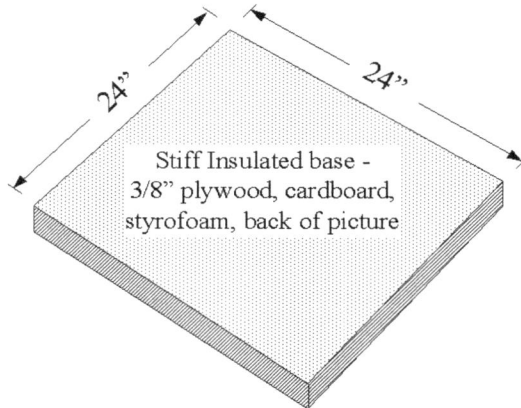

SUBSTRATE FOR THE HORIZONTAL LOOP ANTENNA
FIGURE 10-1

Cut a 6" wide piece of aluminum foil. Lay it on the base, with its edge along the edge of the base. Note the 6 1/2" gap where the foil begins.

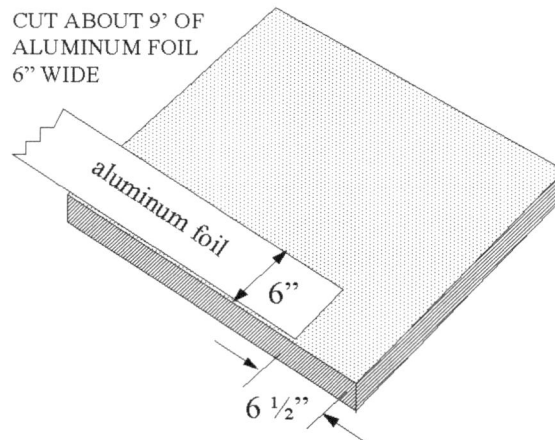

STARTING THE ANTENNA
FIGURE 10-2

Fold the foil over itself. Use tabs of scotch tape to hold the foil in place. Continue taping it down on three, and part of the fourth, side.

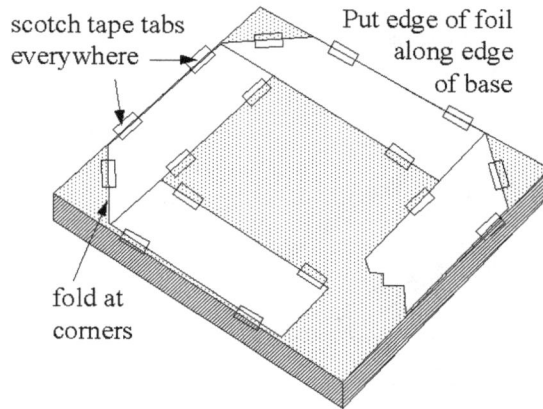

scotch tape tabs everywhere

Put edge of foil along edge of base

fold at corners

CONTINUING TO BUILD UP THE ANTENNA
FIGURE 10-3

Cut a piece of newspaper 7" wide and 20" long. You can use saran wrap, waxed paper, or lots of other thin, insulated material for this - even a handkerchief! But thick insulation could prevent tuning the antenna to low HF frequencies. Put the insulating sheet over the first part of the foil and tape it in place.

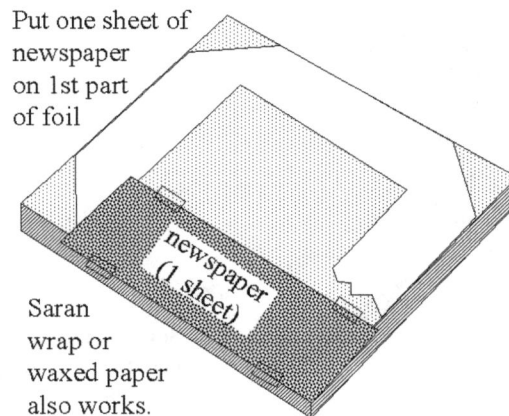

Put one sheet of newspaper on 1st part of foil

newspaper (1 sheet)

Saran wrap or waxed paper also works.

ADDING THE "CAPACITOR" INSULATION
FIGURE 10-4

Leave the fifth flap of foil loose, so it can be lifted up when necessary. Cut off excess foil so it doesn't short circuit against the foil beneath the newspaper.

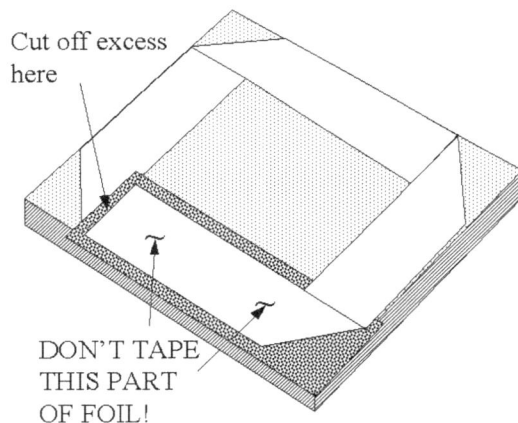

Cut off excess here

DON'T TAPE THIS PART OF FOIL!

FINISHING THE TUNING CAPACITOR UPPER PLATE
FIGURE 10-5

Set the antenna on a table, couch, bed, or on the floor. Keep it horizontal. Put a portable receiver on the foil as shown. The capacity between the receiver and the foil effectively "grounds" the receiver. Pull the whip antenna out, and lay it on the other side of the foil loop. Use a small weight to press the whip against the foil.

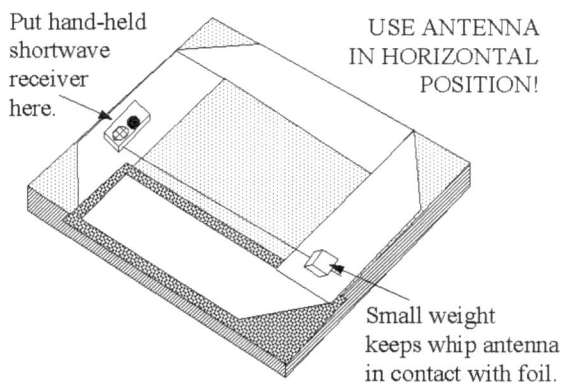

Put hand-held shortwave receiver here.

USE ANTENNA IN HORIZONTAL POSITION!

Small weight keeps whip antenna in contact with foil.

CONNECTING A POCKET-STYLE SHORTWAVE RECEIVER
FIGURE 10-6

Use a book to press the loose foil flap down against the newspaper. Pull excess foil up and back over the book. The book will effect the antenna tuner.

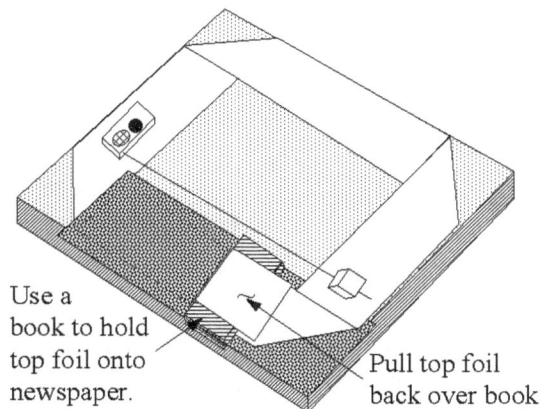

Use a book to hold top foil onto newspaper.

Pull top foil back over book

COMPLETING THE ANTENNA TUNER (TUNING CAPACITOR)
FIGURE 10-7

Tune the antenna to the right frequency by sliding the book back and forth while keeping excess foil up and over the book. While tuning, set the receiver about the middle of the desired band, with the volume control fairly high. Tune for strongest signal, or strong increase in noise from the receiver.

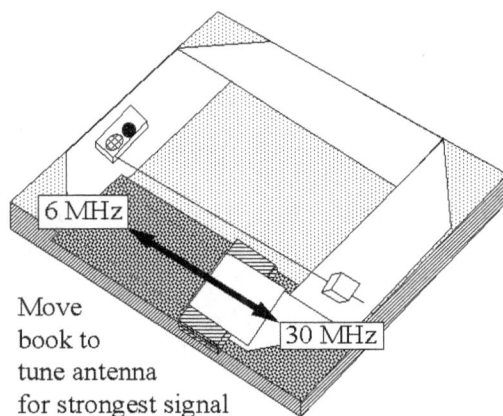

6 MHz

Move book to tune antenna for strongest signal

30 MHz

TUNING THE ANTENNA
FIGURE 10-8

If you use a table model receiver (or an amateur transceiver) make a single turn loop of about #18 insulated hookup wire by taping the wire in place one inch inside the foil loop as shown. Bring the wire away from the antenna on the middle of the side opposite the tuner. Don't connect this loop to the foil - it isn't necessary.

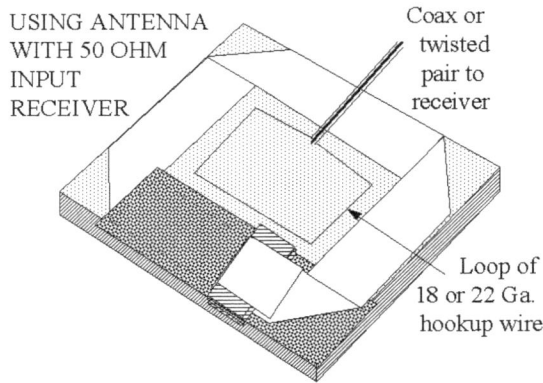

ADDING A PICKUP LOOP TO FEED A RECEIVER WITH COAX INPUT
FIGURE 10-9

The pickup loop can be brought to the receiver by twisted pair or by RG-58 or similar coaxial cable.

CAUTION

THIS ANTENNA IS NOT DESIGNED FOR TRANSMITTING!

Now here's the really neat part of Mike's design - how to cancel noise. Put the antenna on some insulating support so it can be tipped up on one side or corner. One of the early investigators working with us (Cheryl Hagn) pointed out that a pillow serves very well for this support! While listening to the interfering noise, tip the antenna a bit to reduce the noise. Many times, with just a minute or so of adjusting, the noise from power lines, nearby TV sets, etc. can be reduced 20 dB, and further noise canceling can be obtained with a bit more care. See figure 10-10 for Cheryl's way to adjust the antenna for best signal to noise ratio (minimum noise).

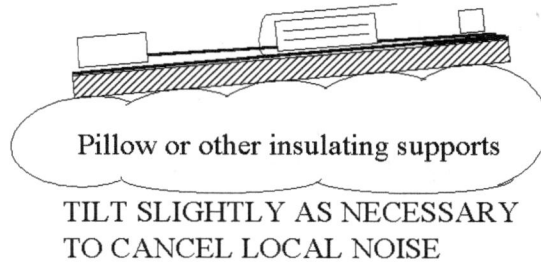

Pillow or other insulating supports

TILT SLIGHTLY AS NECESSARY
TO CANCEL LOCAL NOISE

**TO CANCEL NOISE, KEEP THE ANTENNA ROUGHLY
HORIZONTAL. TIP IT SLIGHTLY TO REDUCE NOISE PICKUP.
Figure 10-10**

Great! You now own the "HLA" antenna for noise canceling. I finish with a
simple schematic diagram showing what you built.

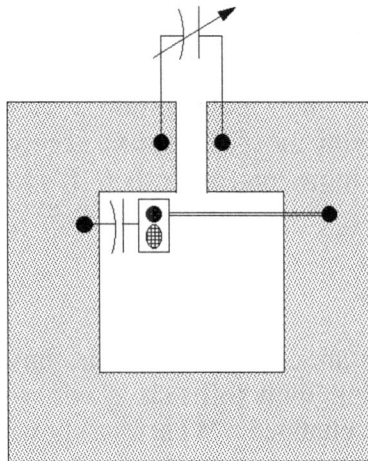

THE TUNED WIDE STRIP LOOP

**SCHEMATIC - VILLARD'S "HORIZONTAL LOOP ANTENNA"
THE "HLA"
FIGURE 10-11**

The newspaper and two foil flaps make a capacitor. The book allows you to vary the amount of overlap of foil, so you made a variable capacitor which is attached to the ends of a single turn loop, itself made of a very wide, flat conductor.

The receiver, sitting on the foil, has a good deal of capacity between its internal circuit board (or chassis) to one end of the loop. Its whip antenna is connected to the other end, so the voltage developed across the loop is injected into the receiver.

Tuning the loop causes the voltage at one frequency to be maximized. This causes the increase in signal strength.

Why the Antenna Rejects Noise

Noise from local sources - arcing power lines, fluorescent tubes, TV sets, etc - travels to your receiver along the ground. Its horizontally polarized component is attenuated very rapidly, so only its vertically polarized part gets through. See chapter 5 on reflection coefficients to appreciate this.

The HLA is horizontally polarized. Therefore it doesn't respond to the noise. But skywave signals arrive at your receiver randomly polarized, so their horizontal part enters the antenna. As their polarization varies, though, the signal will "fade." But this happens with most shortwave receiving antennas. So Mike's HLA suffers from fading no more than any other. It's forte is noise cancellation.

Some Additional Thoughts

There's really nothing magic about the dimensions shown. HLA antennas can be made both larger and smaller, with foil that is wider or narrower. The tuning range will vary if the size is changed. What I've shown has been built in my lab and works from 6 to 30 MHz.

If you decide to make a bigger HLA, just keep in mind that the total circumference of the outside edge of conductor should be kept to well under one-third wavelength. Otherwise, the result will no longer be a small, horizontal loop above ground, but will have other (maybe even interesting) properties.

We've made HLA's from material other than aluminum foil. Sheet aluminum, sheet steel, window screen wire (but not fiberglass, which was the reason for my first failure to get one to work!) or lots of other conducting material will work fine. When using strips of metal, Cheryl discovered you don't need to electrically bond the pieces together. She just put weights at the corners, and the capacity through the oxide or paint layers was essentially a very good connection at RF frequencies.

The HLA was designed to cancel vertically polarized noise. To do so, it should be kept within about a tenth of a wavelength above ground - not elevated much at all. You can, of course, put them up much higher, but I offer no data on performance when the loop is elevated - other people can attest to the qualities of elevated, horizontal loop antennas.

I hope you find this antenna interesting and useful. Some time ago, I had built one and was using it for reception on 15 meters, while using a Grasswire (see chapter 7) for transmitting. In contact with a ham in Europe, I noticed the lights blinking on and off slightly. When I quit with the QSO, I went upstairs to ask my XYL what the heck she was doing - only to find myself in the middle of a raging thunderstorm! The Grasswire, being rather impervious to lightning, and the HLA canceling QRN from the lightning, kept me from realizing that a storm was even in progress!

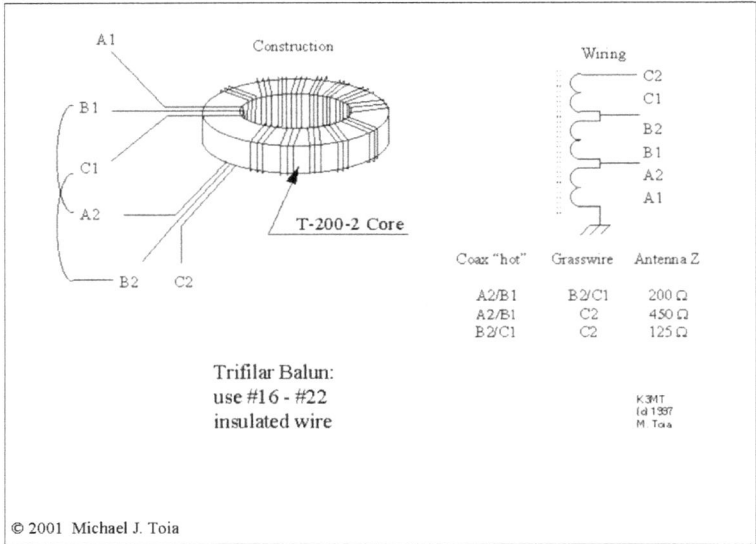

Construction

Winng

A1
B1
C1
A2
B2   C2

T-200-2 Core

C2
C1
B2
B1
A2
A1

Coax "hot"	Grasswire	Antenna Z
A2/B1	B2/C1	200 Ω
A2/B1	C2	450 Ω
B2/C1	C2	125 Ω

Trifilar Balun:
use #16 - #22
insulated wire

K3MT
(c) 1997
M. Toia

A TRIFILAR (3-WINDING) BALUN
USED FOR THE GRASSWIRE (CHAPTER 7)
AND FOR THE GEEVEE
WINDINGS ARE 24 TURNS OF THREE PARALLEL WIRES
FIGURE 11-1

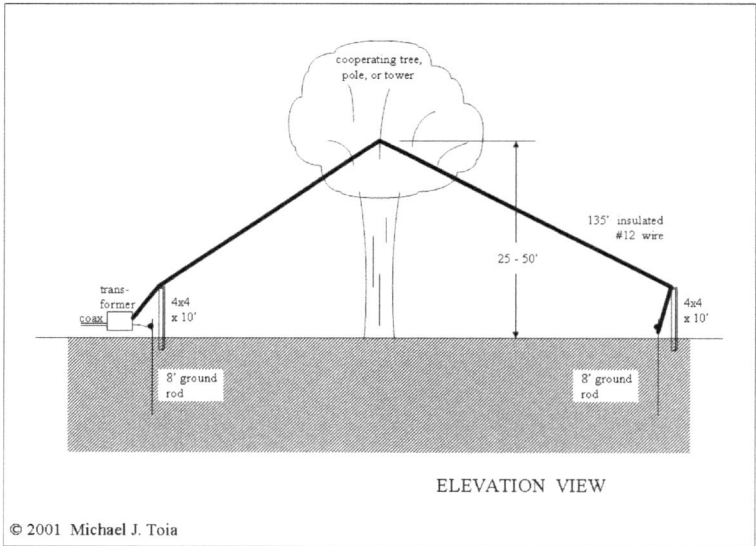

cooperating tree,
pole, or tower

135' insulated
#12 wire

25 - 50'

trans-
former
coax

4x4
x 10'

4x4
x 10'

8' ground
rod

8' ground
rod

ELEVATION VIEW

THE ORIGINAL GEEVEE
FIGURE 11-2

- 11 -

GEEVEE

A GROUNDED INVERTED VEE

This antenna is a variation of the inverted vee. It began as an end-fed, as opposed to center-fed, inverted vee for 80 meters. A half wave dipole for 3.75 MHz would be about 125 feet, end to end. Since I operate CW predominantly, I lengthened the wire to 135 feet. Following usual procedures for installing an inverted vee, I pulled the antenna's center about 40 feet into the air with a halyard, with the help of a cooperating oak tree. There's no balun at this point: the two halves of the dipole are connected together. A 135' continuous piece of wire will do.

The far end was attached, via an insulator and a piece of rope, to a short 4x4 pole, about 6' above ground. The near end (closest to the house) ran over another 4x4 pole, and down to ground level. A standard 8' electrical grounding rod, available at any electrical supply shop and many hardware stores, served as the ground connection at the feed point. Because the antenna is a half-wave on 80 meters, and a multiple of a half-wave on 40, 20, 15, and 10 meters, it is a high voltage, low current, end-fed wire. Therefore not much current need be "sunk" into a counterpoise, so this simple ground system does the job.

A "balun," operating as an unbalanced-to-unbalanced impedance-matching transformer, couples coax to the antenna. See Figure 11-1, a repeat of Figure 7-2 above. Make the balun by removing the four wires from standard household telephone wire. Throw one away, and wind 24 turns (as opposed to 16 turns shown in Chapter 7) of the other three on a T-200-2 Amidon core, as if it were ribbon cable (keep the wires parallel). Connect the three windings in series as shown in the figure to make an RF autotransformer. The coax shield and one end of the balun (A1 in the figure) connect to the ground stake. Connect the antenna to either B2/C1 or to C2, and the coax center conductor to either A2/B1 or B2/C1 in the figure. [13]

---

[13] Very much like feeding the Grasswire antenna.

The antenna matched well, according to the following table of connections:

Frequency, kHz	Coax Center	Antenna
3500	A2/B1	B2
3700	A2/B1	B2
4000	A2/B1	B2
4000	A2/B1	B2/C1
7000	B2/C1	C2
7300	B2/C1	C2
10100	A2/B1	C2
10150	A2/B1	C2
14000	A2/B1	C2
14350	A2/B1	C2
21000	B2/C1	C2
21450	B2/C1	C2
28000	B2/C1	C2
29700	B2/C1	C2

This arrangement worked well when I first tried it. Then I wondered if the end-fed, inverted vee might work somewhat like a Grasswire, end fed, with the far end grounded. So I drove a second ground stake near the far end of the antenna, and connected the antenna wire to it. Figure 11-2 shows the resulting antenna.

The Geevee also works on 160 meters if the far end is removed from the ground stake and attached to a 135' wire laid on the ground. Best match was with the coax center conductor on A2/B1, and antenna on C2.

Since the Geevee is mounted close to, and electrically connected to, the ground at both ends, NEC-2 had trouble grappling with it. I ran some predicted patterns using more advanced modeling software, NEC-4, using W7EL's *EZNEC*. First are 3-D patterns, looking straight down from above, along the negative z axis. The antenna orientation is along the x-axis, with the feed point at negative x (to the right in the patterns.) Think of the antenna running from East to West, fed at the eastern end. The y axis is North, the x axis West.

The 3-D patterns show azimuth patterns for ascending values of takeoff angle, in 5° increments. A 2-D elevation pattern follows each 3-D pattern. This is a cut through the solid pattern, taken along an azimuth having the largest power gain at 10 to 30 degrees.

The 3-D plots show the tendency for the antenna's main lobe to fall closer and closer to the wire as frequency increases. Another way to say the same thing is that the lobe gets closer to the wire as the wire gets longer, measured in wavelengths or degrees. This is a general characteristic of wire antennas.

Elevation Plot
Azimuth Angle   90.0 deg.
Outer Ring   2.89 dBi

3D Max Gain   2.89 dBi
Slice Max Gain   2.89 dBi @ Elev Angle = 26.0 deg.
Beamwidth   46.3 deg. ; -3dB @ 9.0, 55.3 deg.
Sidelobe Gain   2.89 dBi @ Elev Angle = 155.0 deg.
Front/Sidelobe   0.0 dB

Cursor Elev   26.0 deg.
Gain   2.89 dBi
                 0.0 dBmax

3.5 MHz

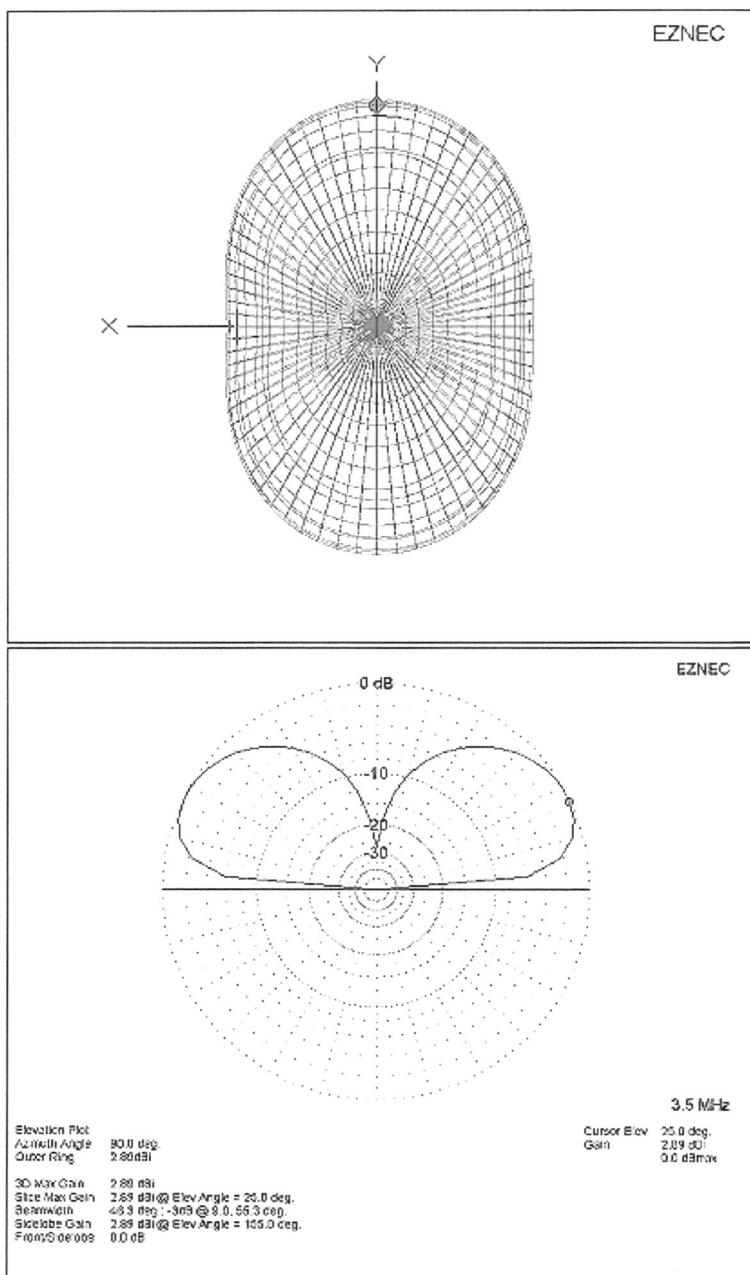

80 METERS
FIGURE 11-3

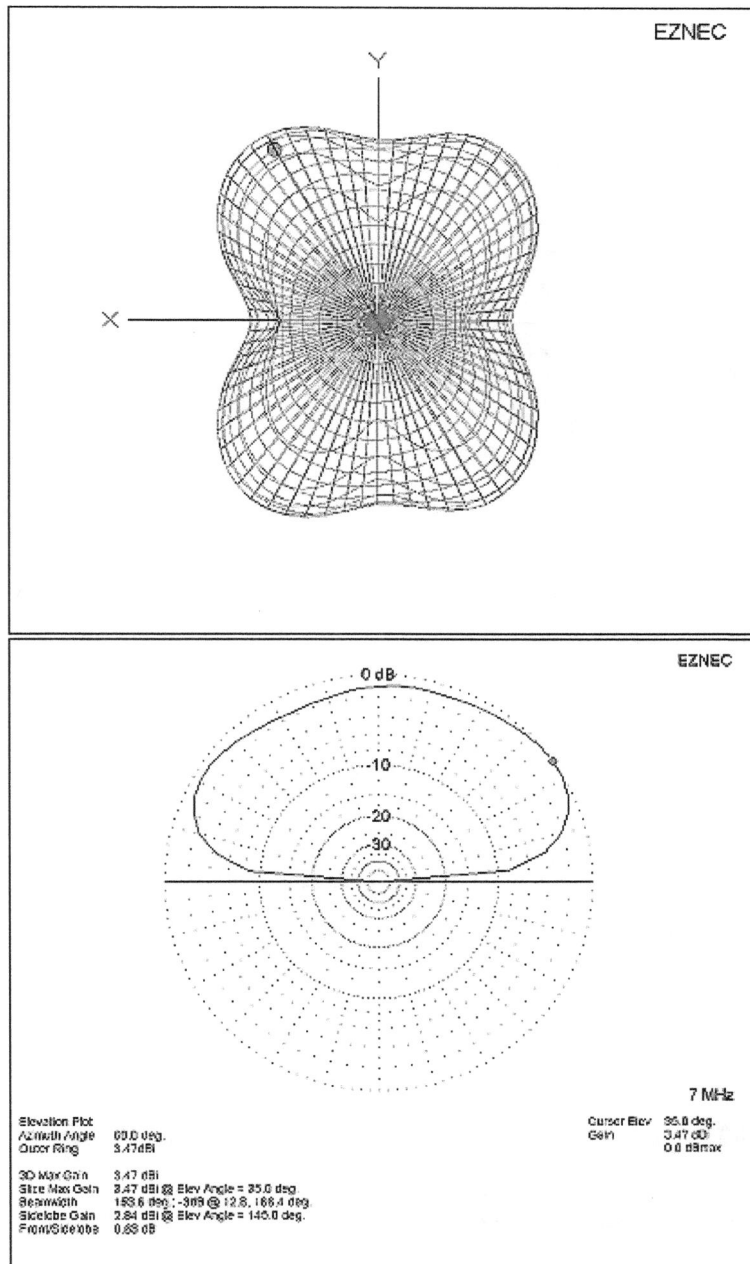

40 METERS
FIGURE 11-4

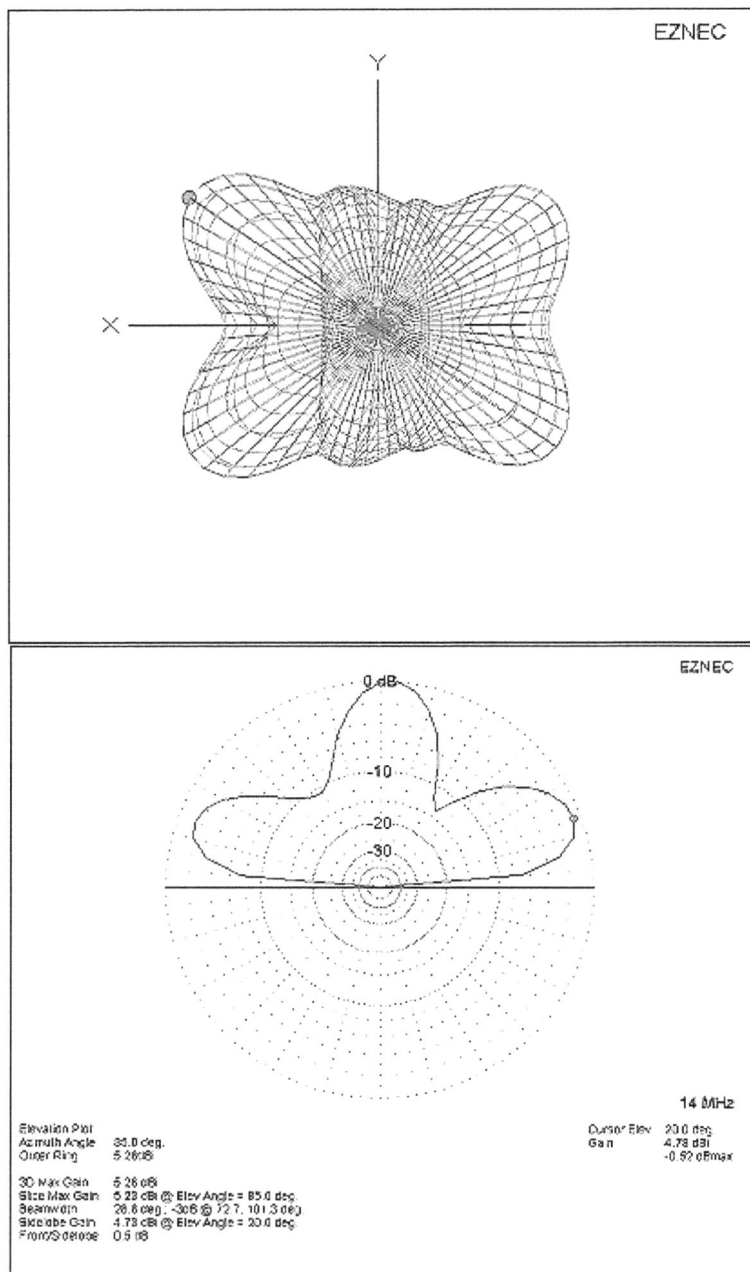

20 METERS
FIGURE 11-5

Elevation Plot
Azimuth Angle   25.0 deg.
Outer Ring   5.51 dBi

3D Max Gain   5.51 dBi
Slice Max Gain   5.46 dBi @ Elev Angle = 16.0 deg.
Beamwidth   19.2 deg.; -3dB @ 7.2, 26.4 deg.
Sidelobe Gain   4.88 dBi @ Elev Angle = 60.0 deg.
Front/Sidelobe   0.57 dB

Cursor Elev   16.0 deg.
Gain   5.46 dBi
              -0.03 dBmax

21 MHz

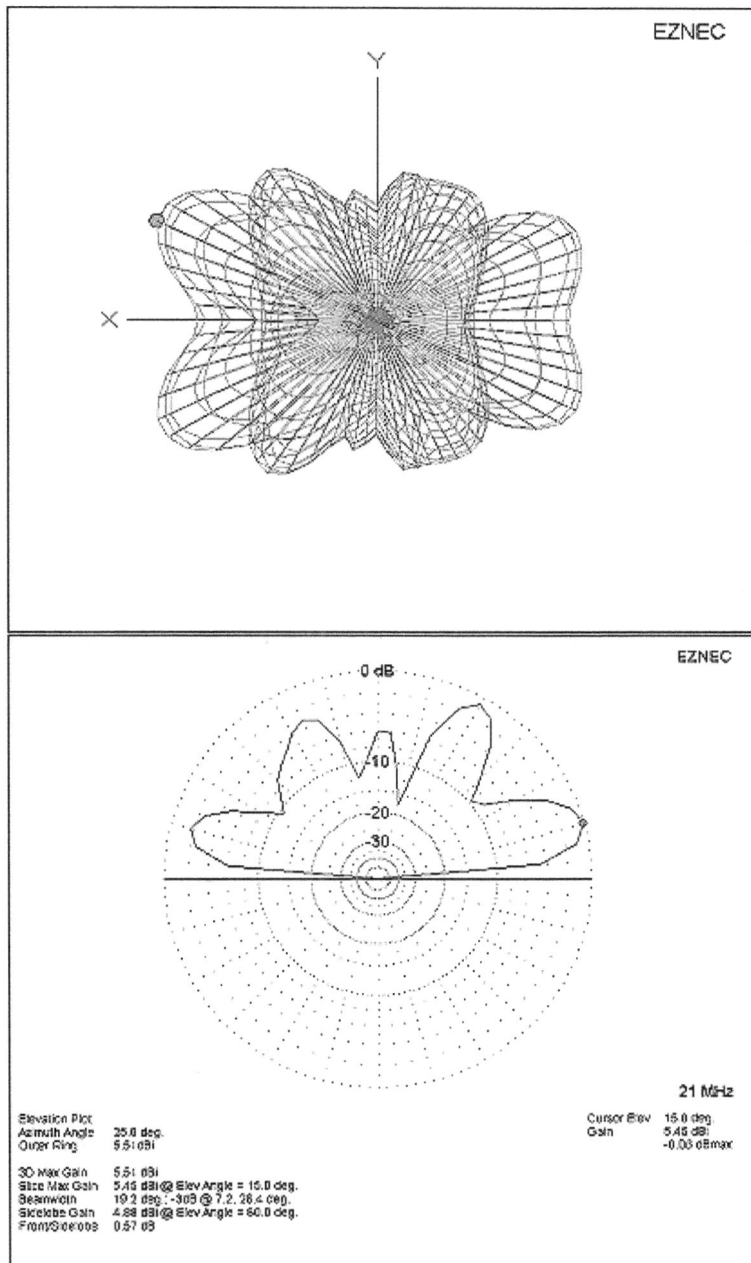

15 METERS
FIGURE 11-6

Elevation Plot
Azimuth Angle   26.0 deg.
Outer Ring   6.83dB

3D Max Gain   6.83 dB
Slice Max Gain   6.08 dB @ Elev Angle = 10.0 deg.
Beamwidth   16.6 deg.; -3dB @ 5.2, 20.0 deg.
Sidelobe Gain   5.52 dBi @ Elev Angle = 45.0 deg.
Front/Sidelobe   0.56 dB

Cursor Elev   10.0 deg.
Gain   6.08 dB
              -0.75 dBmax

28 MHz

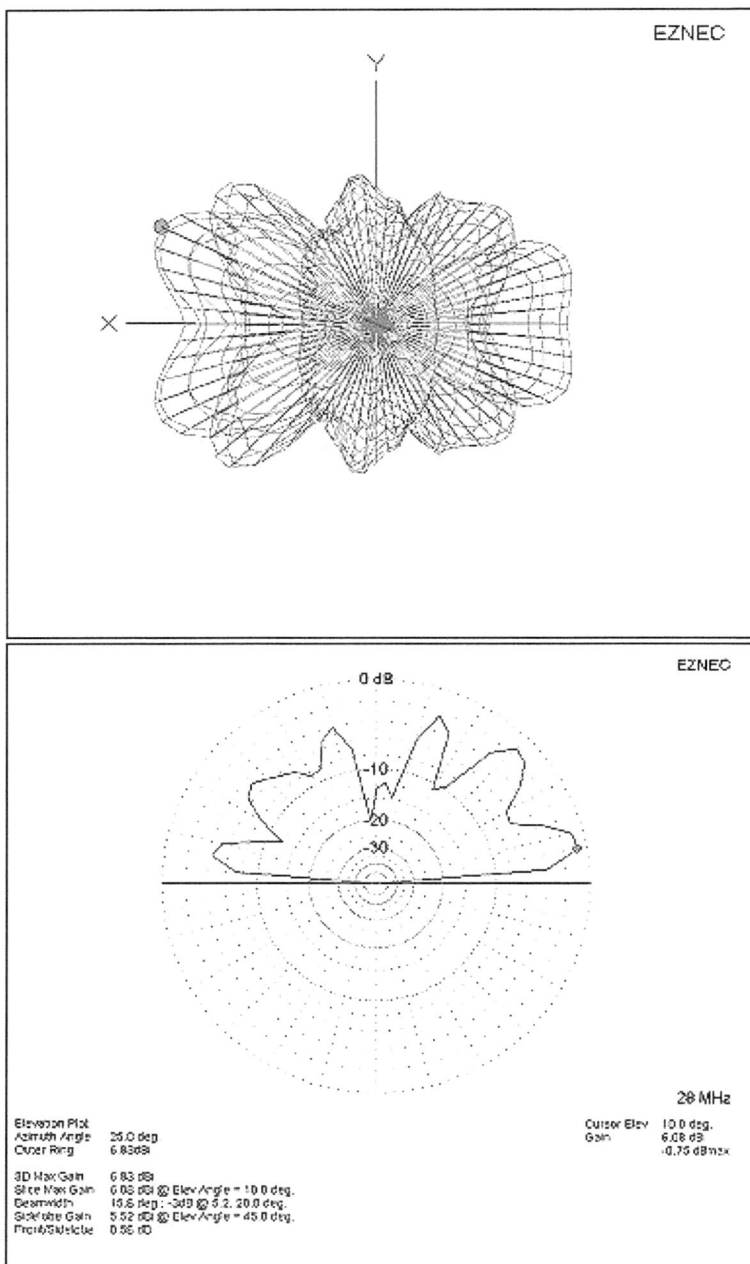

10 METERS
FIGURE 11-7

NOTES

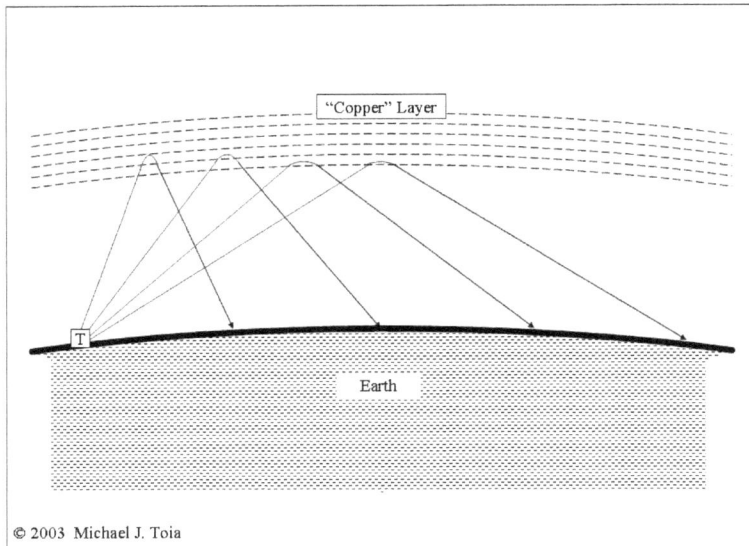

A TOTALLY REFLECTING, "COPPER" IONOSPHERE
FIGURE 12-1

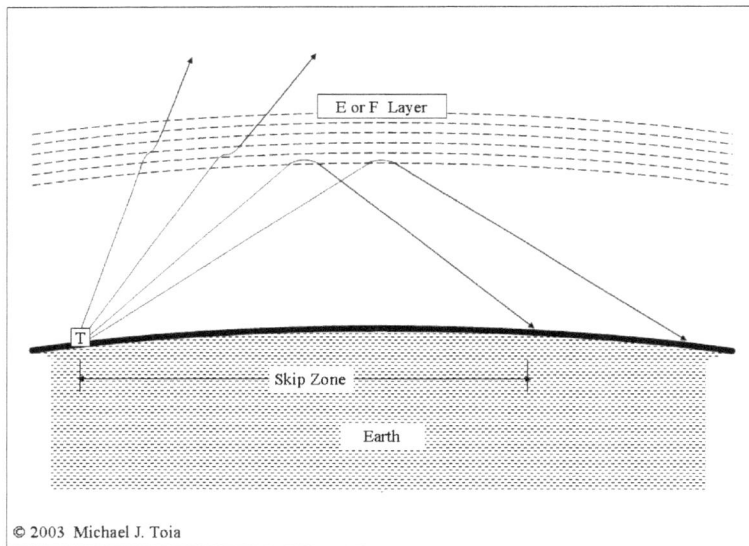

A MORE TYPICAL IONOSPHERE
FIGURE 12-2

89

- 12 -

ANTENNA HEIGHT

AND

TAKEOFF ANGLE

The Case for Low Takeoff Angle

Takeoff angle is the angle between the horizon and the main lobe of radiation leaving your antenna. To work DX, this angle should be low: the lower the better. There's a good reason for this. If a ray of radiation leaving your antenna hits the ionosphere at a glancing angle, it has a better chance of being reflected back to earth than one hitting at too steep an angle. Furthermore, that ray goes farther before hitting the earth again.

The ionosphere is rather complex, and there are a number of computer programs available, some free, that calculate the conditions and band for communication with practically any point on earth. A very simplistic view of the ionosphere is the "copper" model:" any ray that hits it will be reflected. The greater the takeoff angle, the shorter the range, as in Figure 12-1. Good DXing needs a low takeoff angle. For local ragchewing and short-range contacts at HF, a high takeoff angle is required.

There's a "rub" to the copper model. If the ionosphere were a good reflector, it would reflect a ray that hits it at *any* angle, even one traveling straight up. But alas, such a ray (we call *vertical incidence*) is reflected back to its starting point on earth only at frequencies below some maximum, called the *critical frequency*. This varies with solar activity, and is rarely above 10 MHz, if at all. At higher frequencies, a ray launched at too high an angle simply punches through the ionosphere and escapes to outer space, as in Figure 12-2.

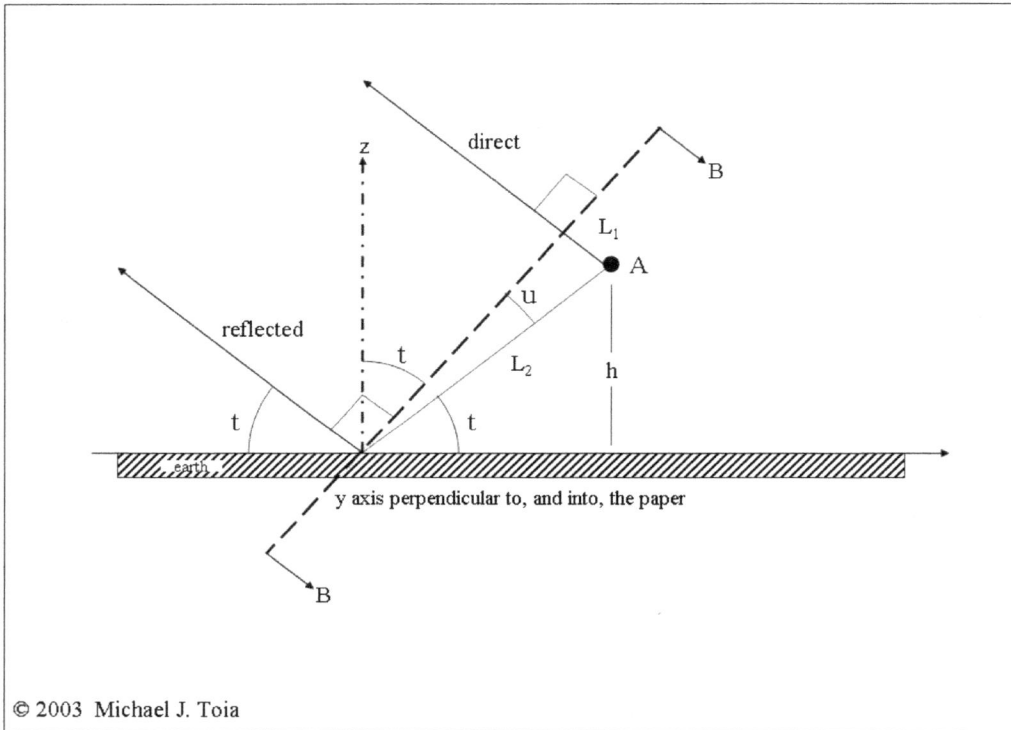

PHASE DELAY, REFLECTED RAY COMPARED TO DIRECT RAY
FIGURE 12-3

There is often a zone from the transmitter, marked "T" on Figures 12-1 and 12-2, where no signal comes back to earth. The signal *skips* over this area, named the "skip zone" for that reason. The ionosphere's refracting power (ability to bend the radio ray) decreases with increasing frequency. Below the critical frequency, no skip zone exists. As the frequency increases more and more above the critical frequency, the skip zone gets longer and longer. That's why local contacts, out to distances of a few hundred miles, are easy on 75 meters and 40 meters, but not nearly as common on 20 meters or higher bands.

Takeoff Angle vs. Height

In chapter 5 I noted that the earth reflects radio waves differently for horizontal and vertical polarization. The strength of the reflected ray and its phase angle change, depending on the takeoff angle. A bit of common mathematics is involved to see if the reflected ray adds to, subtracts from, and is strong enough to influence, the antenna's direct ray. Modern computer programs do the math so nicely, making this job easier.

For horizontal polarization, the reflected ray is almost as strong as, but phase shifted almost 180° from, the direct ray. For an antenna close to the ground, these two rays cancel, producing little radiation. When the round trip of the reflected ray, from the antenna to the ground and back, is 180°, the two <u>add</u>, doubling the electric field strength for a 6 dB increase in gain over free space conditions. This happens when a horizontal dipole is a quarter wavelength or higher above ground. For vertical polarization, the situation is more complex, because the reflection coefficient goes through such a change in both strength and phase shift

Figure 12-3 shows that an antenna launches two rays toward the ionosphere. The direct ray travels a distance L_1, and the reflected one, L_2, to reach a plane perpendicular to the outgoing wavefront. Hence the reflected ray is delayed by having to travel the greater distance, $L_2 - L_1$. This, along with any phase shift caused by the reflection coefficient, brings the two into, or out of, phase, causing partial cancellation or addition of the two.

I used NEC-2 - calculated takeoff angle patterns of short dipoles mounted above ground, horizontally and vertically polarized, and a log periodic horizontal beam at higher frequencies. The figures that follow are the results for heights of 15, 30, 45, 60, and 75 feet, as indicated.

In all cases, the left figure was calculated for good soil conditions, dielectric constant = 20, and conductivity = 20 millisiemens/meter (a.k.a. mmho/m). The right figure was calculated for poor soil, dielectric constant = 6, and conductivity = 1 millisiemens/meter The patterns appear in four groups, two for horizontal polarization, and two for vertical. Each group occupies two facing pages. The software plots zenith angle, not takeoff angle. Each pattern runs from zero degrees (90° zenith angle, or "up") to 90 degrees (0° zenith angle, or to the right) takeoff angle. That is, the angle shown is 90° - takeoff angle.

Group 1 shows calculated patterns for a horizontal dipole on 80, 40, 30, and 20 meters. These are bands radio amateurs are most likely to use dipole antennas, although they are also used on higher bands. Patterns are on a scale of -10 to +10 dBi.

Group 2 shows patterns for a modest horizontally polarized beam antenna on 20, 15, and 10 meters, bands where most beams are used. Again some amateurs use yagis on 40 and, yes, even 80 meters, but these are rare. Patterns for dipoles can be inferred from the beam patterns by reducing overall gain by about 6 dB. The scale for beam patterns is -4 to +16 dBi.

Group 3 shows patterns for vertical dipoles on 80, 40, and 30 meters. Patterns for elevated ground plane antennas will be similar in shape. The patterns span -14 to +6 dBi: vertical dipoles simply do not have the same gain as horizontal dipoles for reasons explained earlier.

Group 4 shows patterns for vertical dipoles on 20, 15, and 10 meters, with the same -14 to +6 dBi scale as used for lower frequency verticals.

There are some curios effects indicated in the vertical patterns, particularly for higher frequencies at higher elevations. It seems that, when mounted over poor soil, the vertical will do better at a low takeoff angle than when mounted over good soil. I let the reader draw what conclusions he/she may wish about this effect.

Calculated patterns under differing conditions follow.

Takeoff angle patterns for a horizontal dipole on 80 to 20 meters. Gain scales run from -10 to +10 dBi. The angle scale is in terms of zenith angle: 0° is directly overhead, while 90° is along the horizon. Takeoff angle = 90° - zenith angle.

Patterns on the left are for good soil conditions typical of well-irrigated, wet farm land. Patterns on the right are for poor soil conditions typical of fairly dry and/or deep sandy soils. Note that deep nulls in the pattern occur for good soil conditions, but nulls are less distinct over poor soils.

Patterns clearly show the optimum height for close-in and DX communication. For 40 meters, 30' is a good height for local ragchewing, while 60' is not nearly as good. On the other hand, higher is better for DXing.

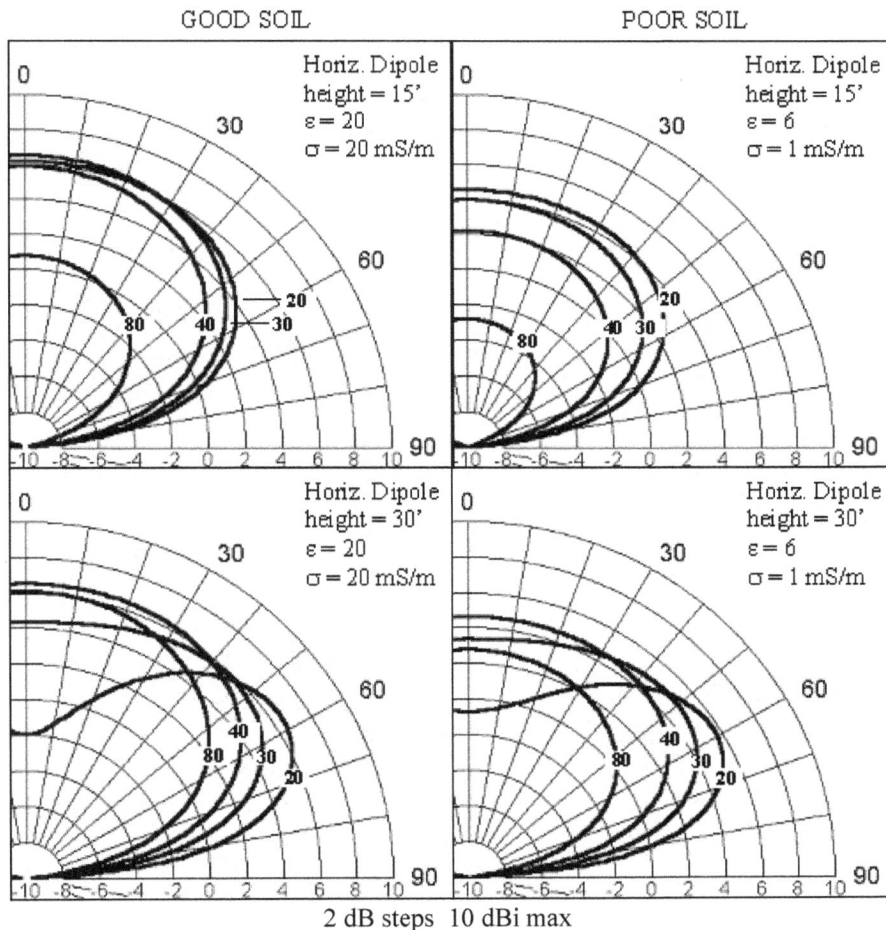

FIGURE 12-4

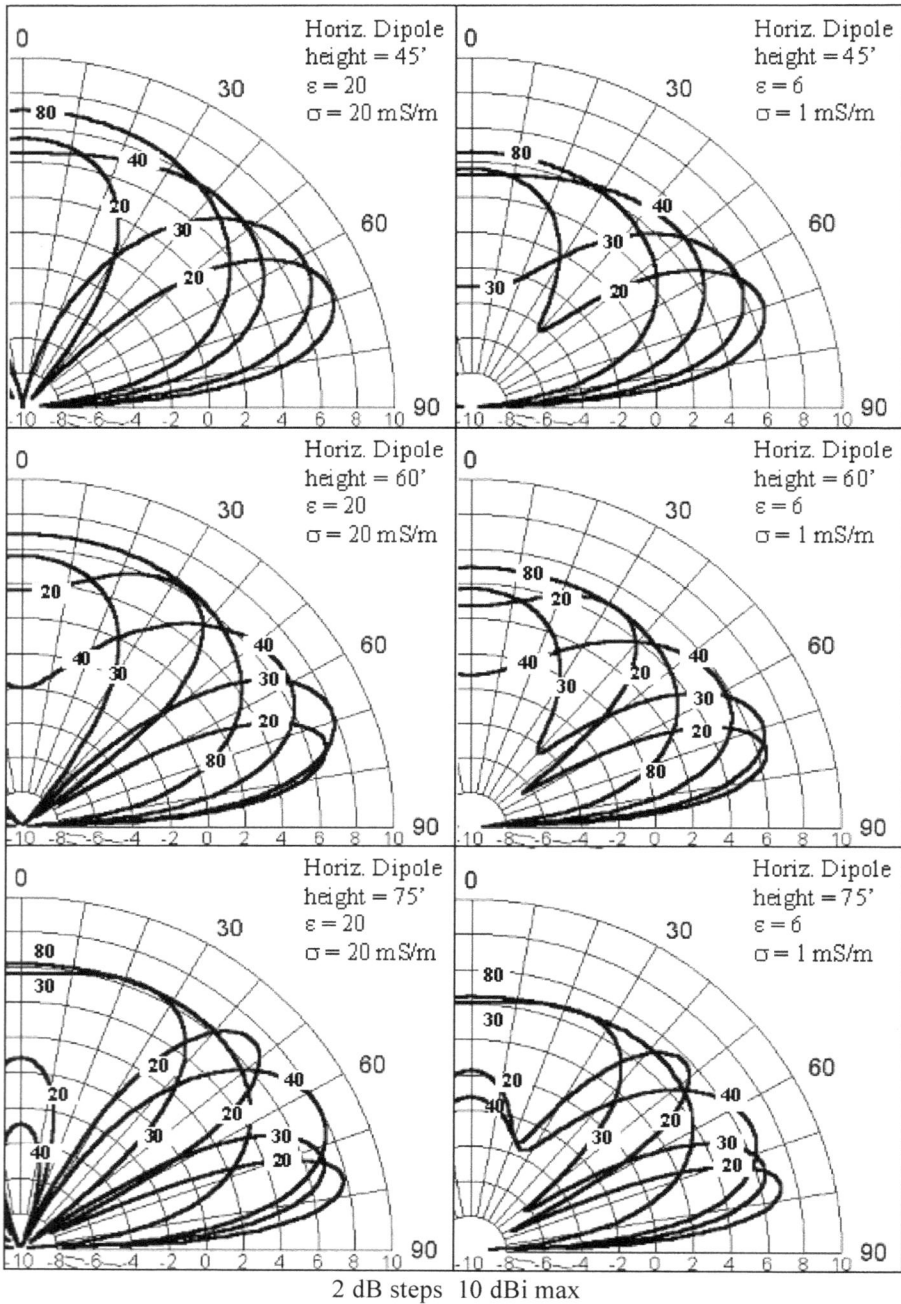

Horiz. Dipole
height = 45'
ε = 20
σ = 20 mS/m

Horiz. Dipole
height = 45'
ε = 6
σ = 1 mS/m

Horiz. Dipole
height = 60'
ε = 20
σ = 20 mS/m

Horiz. Dipole
height = 60'
ε = 6
σ = 1 mS/m

Horiz. Dipole
height = 75'
ε = 20
σ = 20 mS/m

Horiz. Dipole
height = 75'
ε = 6
σ = 1 mS/m

2 dB steps   10 dBi max

FIGURE 12-5

95

These are patterns for a horizontal beam antenna on 20, 15, and 10 meters. Gain scales run from -4 to +16 dBi. Again, the angle scale is in terms of zenith angle: 0° is directly overhead while 90° is along the horizon. Takeoff angle = 90° - zenith angle. Patterns for 17 meters can be interpolated between 20 and 15 meters. For 12 meters, interpolate between 15 and 10 meters.

Patterns on the left are for good soil conditions typical of well-irrigated, wet farm land. Patterns on the right are for poor soil conditions typical of fairly dry and/or deep sandy soils. Note that deep nulls in the pattern occur for good soil conditions, but nulls are less distinct over poor soils.

Patterns clearly show the optimum height for DX communication. Higher is generally better. A half wave height is about optimum.

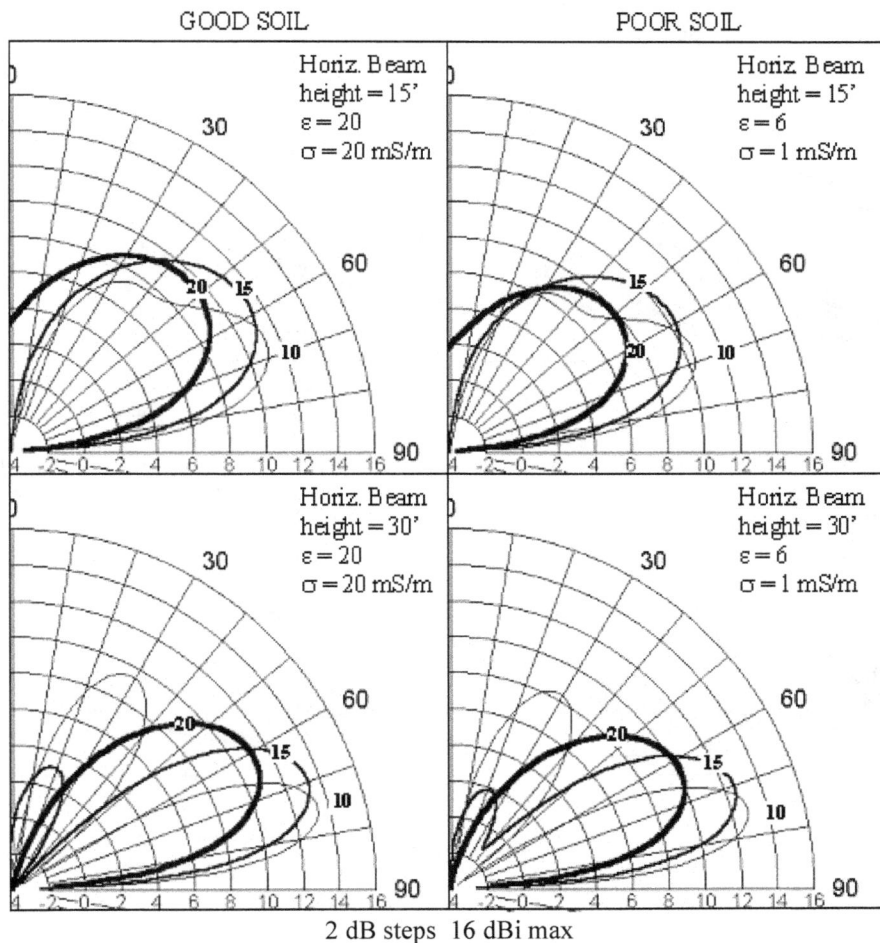

2 dB steps 16 dBi max

FIGURE 12-6

96

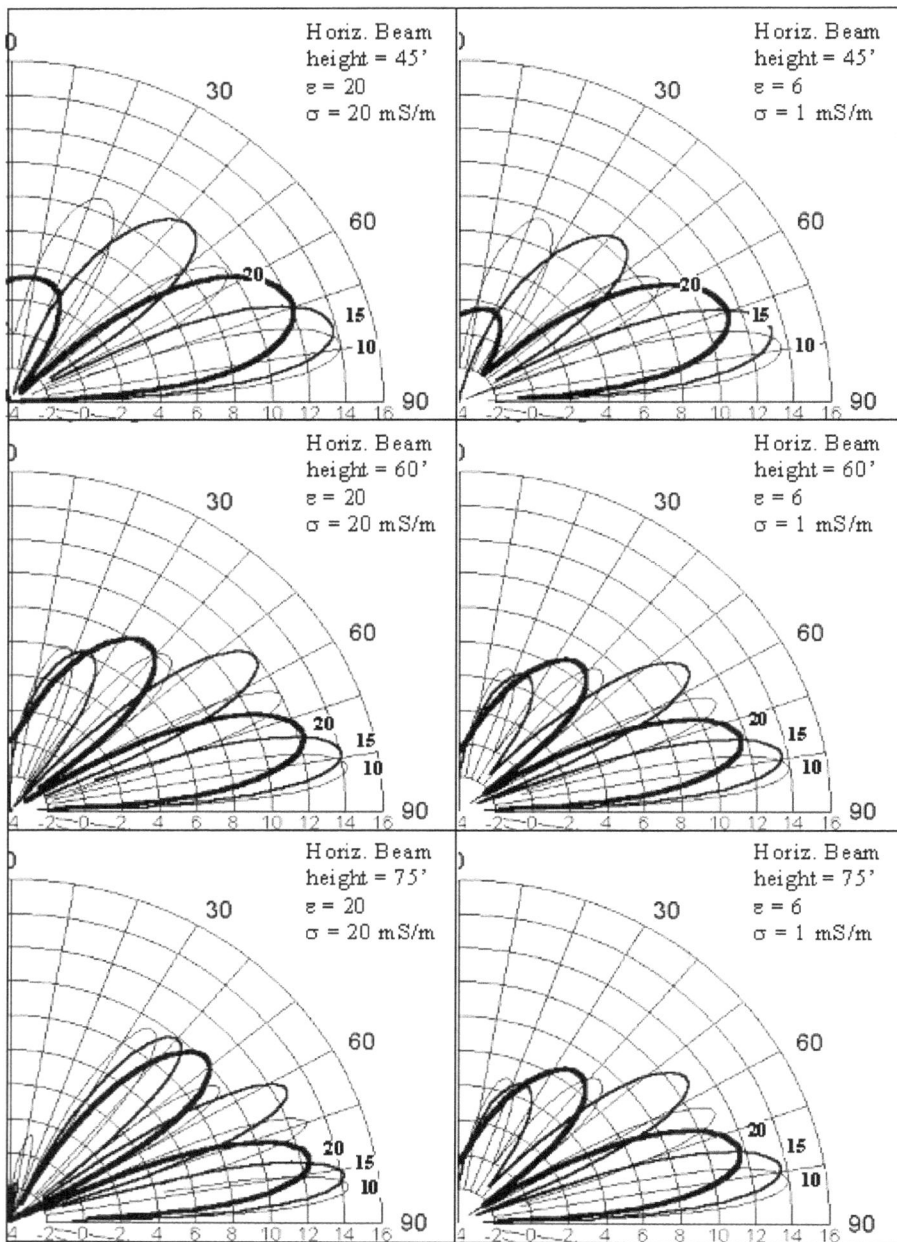

Horiz. Beam height = 45' ε = 20 σ = 20 mS/m

Horiz. Beam height = 45' ε = 6 σ = 1 mS/m

Horiz. Beam height = 60' ε = 20 σ = 20 mS/m

Horiz. Beam height = 60' ε = 6 σ = 1 mS/m

Horiz. Beam height = 75' ε = 20 σ = 20 mS/m

Horiz. Beam height = 75' ε = 6 σ = 1 mS/m

2 dB steps 10 dBi max

FIGURE 12-7

These are calculated patterns for a vertical dipole antenna on 80, 40, and 30 meters. Gain scales run from -14 to +6 dBi. The angle scale is again in terms of zenith angle: 0° is directly overhead while 90° is along the horizon.

Patterns on the left are for good soil conditions. Those on the right are for poor soil conditions. On 80 meters, higher is better. On 40 and 30 meters, note that the antenna can be too high, where it launches a lot of energy "up", and not as much for DXing, due to splitting of the main lobe.

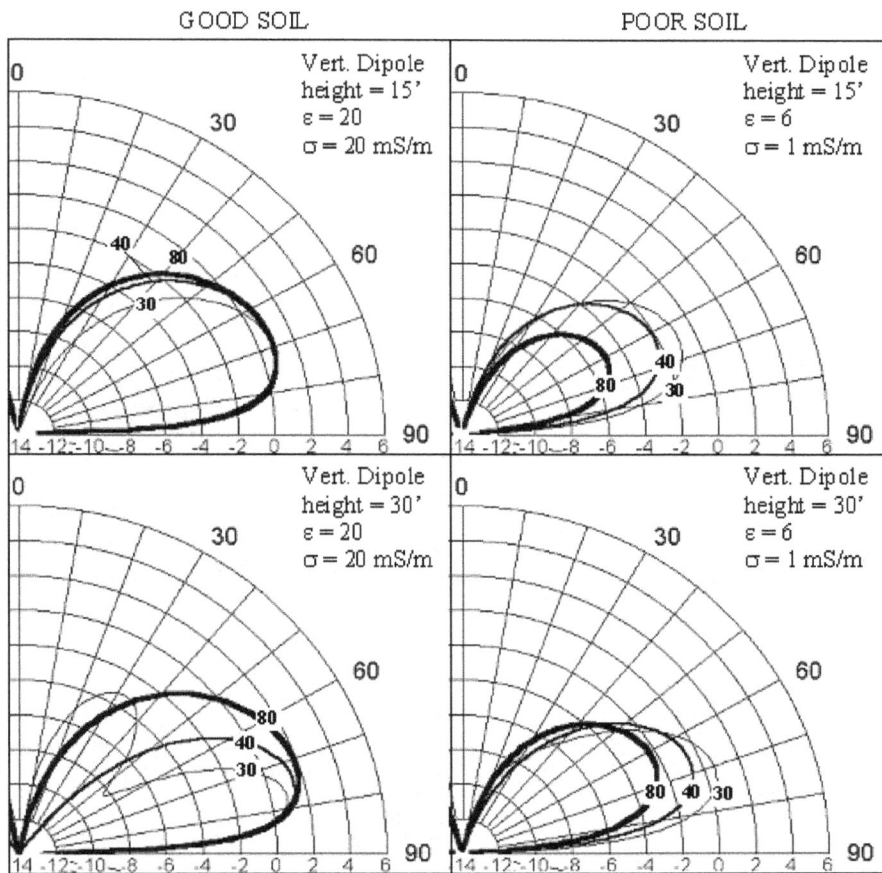

2 dB steps 6 dBi max

FIGURE 12-8

98

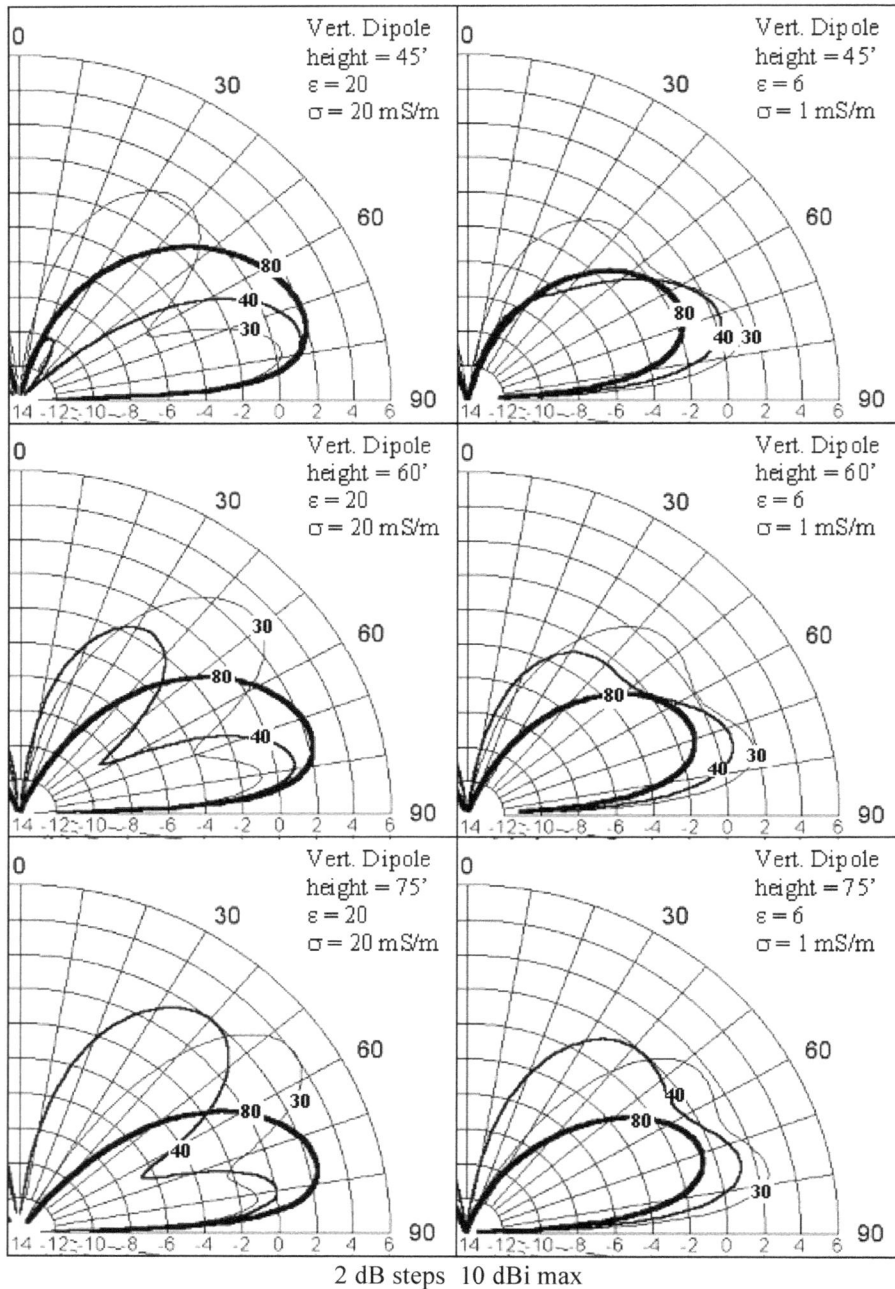

2 dB steps 10 dBi max

FIGURE 12-9

99

Takeoff angle patterns for a vertical dipole antenna on 20, 15, and 10 meters. Gain scales run from -14 to +6 dBi. The angle scale is in terms of zenith angle. 0° is directly overhead while 90° is along the horizon.

Patterns on the left are for good soil conditions. Those on the right are for poor soil.

Higher is better. There is a curious advantage for a vertical over poor soils, on the order of a dB or two. This is related to the reflection coefficient at low takeoff angles, where good soils reflect at close to 180° phase shift, but poor soils reflect closer to 270°. The reflected ray for low takeoff angles tends to cancel radiation, but the additional shift of poor soils actually boosts radiation a slight amount.

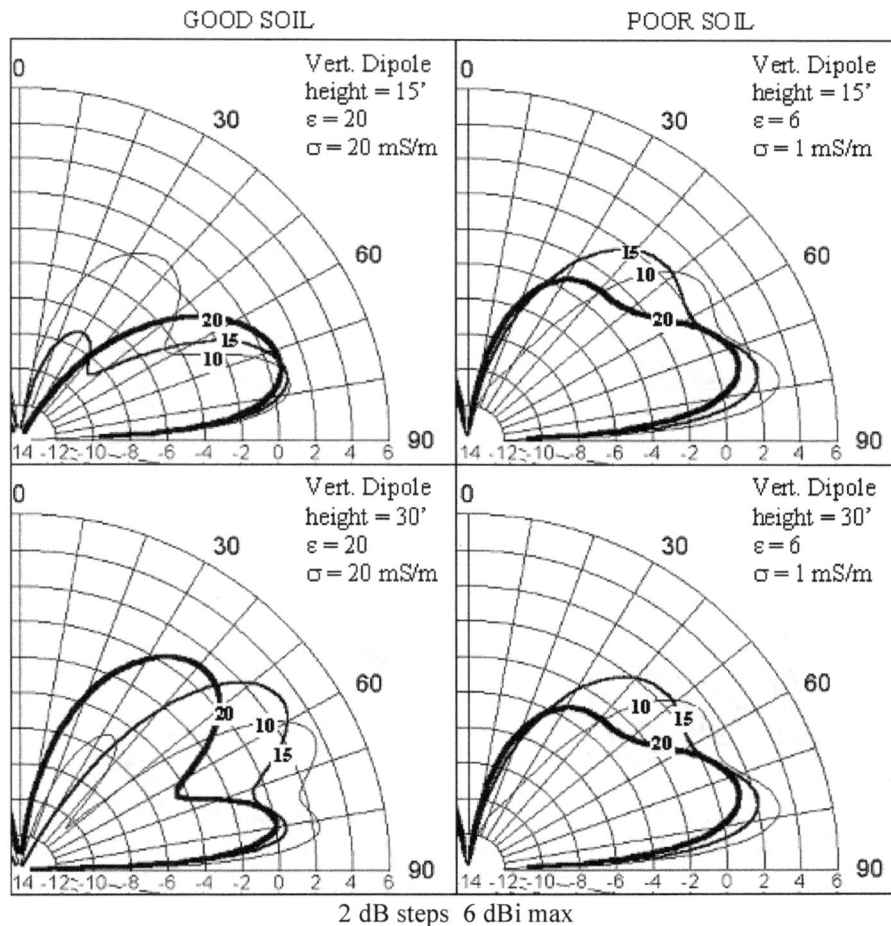

GOOD SOIL   POOR SOIL

Vert. Dipole
height = 15'
ε = 20
σ = 20 mS/m

Vert. Dipole
height = 15'
ε = 6
σ = 1 mS/m

Vert. Dipole
height = 30'
ε = 20
σ = 20 mS/m

Vert. Dipole
height = 30'
ε = 6
σ = 1 mS/m

2 dB steps 6 dBi max

FIGURE 12-10

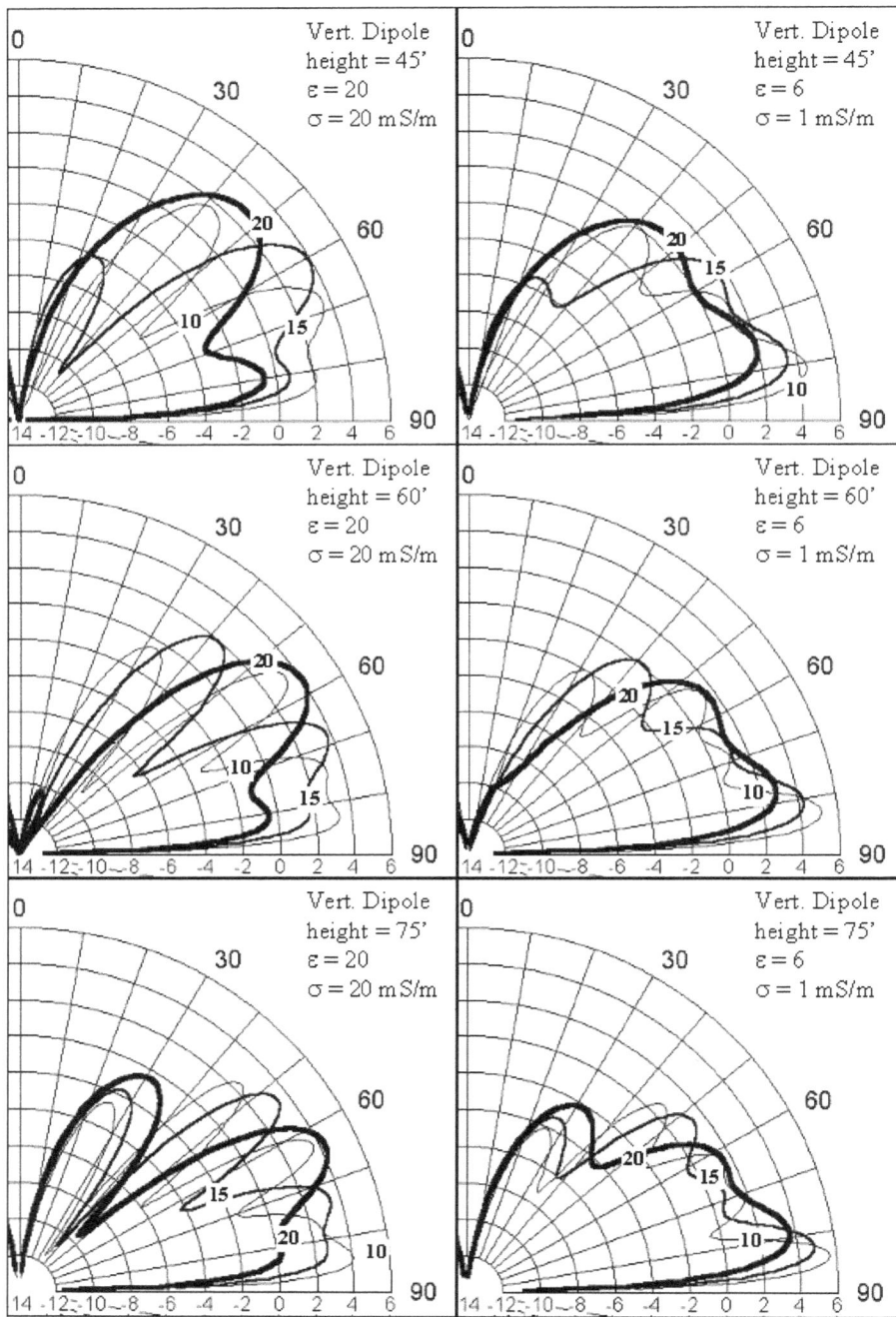

Vert. Dipole height = 45' ε = 20 σ = 20 mS/m

Vert. Dipole height = 45' ε = 6 σ = 1 mS/m

Vert. Dipole height = 60' ε = 20 σ = 20 mS/m

Vert. Dipole height = 60' ε = 6 σ = 1 mS/m

Vert. Dipole height = 75' ε = 20 σ = 20 mS/m

Vert. Dipole height = 75' ε = 6 σ = 1 mS/m

2 dB steps 6 dBi max

FIGURE 12-11

101

NOTES

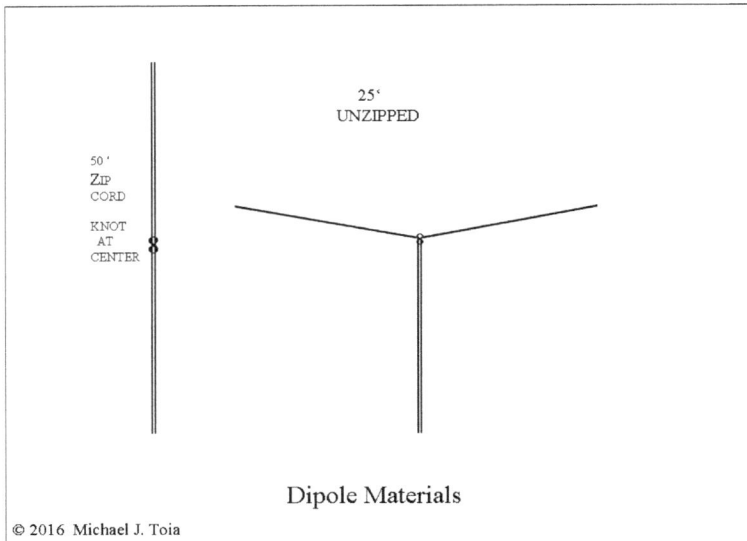

25'
UNZIPPED

50'
ZIP
CORD

KNOT
AT
CENTER

Dipole Materials

ZIP CORD DIPOLE
FIGURE 13-1

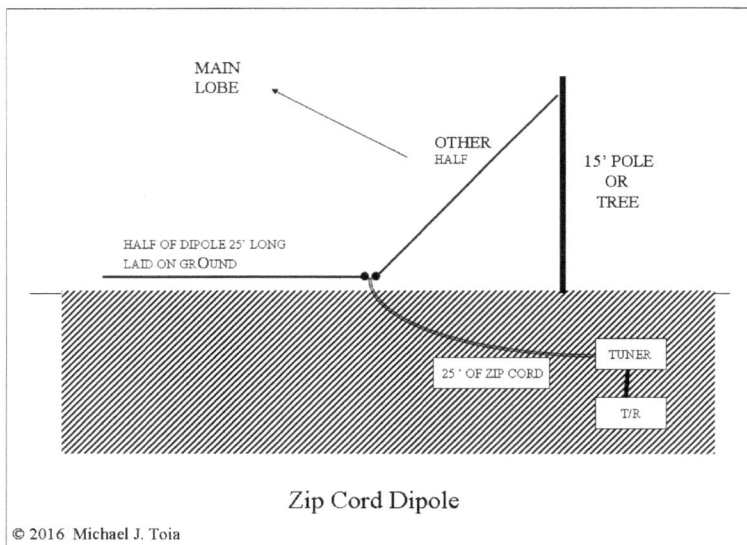

MAIN
LOBE

OTHER
HALF

15' POLE
OR
TREE

HALF OF DIPOLE 25' LONG
LAID ON GROUND

25' OF ZIP CORD

TUNER

T/R

Zip Cord Dipole

DIPOLE DEPLOYMENT
FIGURE 13-2

- 13 -

A QUICK-UP PORTABLE DIPOLE

This write-up is an addition to the first edition. It describes a very simple dipole antenna quite useful for HF communication. The dipole is made of standard electrician's "zip cord" sold in hardware stores and many other places, a two-conductor cord used for wiring lamps and the like. Buy fifty feet of the cord. Tie a knot at its center, then peel the two conductors apart from one end to the knot as in Figure 13-1. You will now have a wire dipole, fifty feet end to end, with a twenty-five foot parallel wire transmission line.

Use a bit of twine and a small weight. Toss the weigh over a cooperating tree, and allow it to fall to the ground. Hoist one end of the dipole up into the tree, about eighteen feet above ground. In lieu of a tree you can use a long, extendable painter's pole.

Run the elevated wire to the ground at about a 45 degree angle, and fasten the knot to a small stake. Then spread the other end of the wire away from the tree. Run the non-split transmission line at about a right angle away from the dipole, and attach it to a tuner. Figure 13-2 shows the completed antenna.

These dipoles can be nicely effective. At 40 meters I have noticed a decided directional property to them. From my home at the time a bit North of Washington DC, I was working a W3 just North of Pittsburgh, an East-to-West path of about 200 miles. Signals were a bit weak as the dipole was oriented North-South. I asked the other chap to standby a bit while I "turned the beam." With that, I pulled up stakes and oriented the dipole with its sloping end leaning to the East, away from the great circle path. Signals improved one S-unit or more!

GLOSSARY

dBi - Gain of an antenna, in the direction specified, usually the direction of maximum gain. Gain is referenced to a mythical "isotropic" antenna that radiates equally well in all directions when mounted in outer space. 3 dB = double power strength. 10 dB = 10 times power strength. The subscript, "i" says the reference power strength is the isotropic antenna.

degrees - when used as a unit of length, equal to 1/360th of a wavelength. At 3.75 MHz, one wavelength is 80 meters. Thus 360 *degrees* = 80 meters *at this frequency*! 1 meter = 4.5 degrees, and 1 degree = 22.222 . . . cm. At 40 meters, a degree is *half* of this, etc. A half-wave dipole is 180 degrees long, and a quarter wave vertical is 90 degrees tall.

DXer, DX - A ham who has QSO's with hams in distant places, such as Tahiti, France, Australia, etc. *DX* is the other ham. If this is new to you, are you in for a thrill!

feed point - Point where the feedline, often coax, attaches to the antenna.

ionosphere - if you are licensed for HF operation, you should already know what this is. If not, consult the *Radio Amateurs Handbook* or many other references for an explanation. Briefly, it's a mirror made of plasma ("smoke") about 110 to 450 km above the earth. There are three useful layers of this plasma, called the E, F1, and F2 layers. HF propagation is not smoke and mirrors - it's smoke *as* mirrors.

input impedance - The impedance measured at the point where power is to be applied to an antenna. In many cases, the antenna is designed to present 50 ohms of resistance, with low reactance, as its input impedance. This forms a good match to standard coaxial cable.

ladder line - Parallel wire transmission line, somewhat like TV twin lead, but the two wires are kept apart by insulating "sticks" every so often. The appearance is like a stepladder, hence the name.

NEC, NEC-2, NEC-4 - Computer programs that predict performance of antennas. NEC stands for *Numerical Electromagnetic Code*. -2, -3, and -4 are program updates.

In some contexts, could be the *National Electrical Code* to which I refer only in this preface. Otherwise, it means the computer program(s).

reactance - A condition where the (sine wave) current into an antenna is not exactly in phase with the voltage. The current can be divided into the sum of two sine waves, one exactly in phase with the voltage that represents the resistance of the antenna. The other current is 90° out of phase with the voltage. The quotient of the 90° out of phase current divided into the voltage is called *reactance*. If the current peaks before the voltage in time, reactance behaves as if it is a capacitor in series with the antenna resistance. Otherwise, it acts as an inductance: reactance can be *capacitive* or *inductive*.

reflection coefficient - see chapter 5 for a definition. This concerns both how strong the reflection from the ground is, and what phase angle it takes. It sometimes helps, and sometimes hurts.

SWR - The voltage standing wave ratio on the antenna's feedline, generally measured with a SWR bridge, or a SWR meter.

takeoff angle - the angle between a ray leaving the antenna headed toward the ionosphere, and the ground (which is often horizontal, but not always!)

VSWR - See SWR.

www.ingramcontent.com/pod-product-compliance
Lightning Source LLC
Chambersburg PA
CBHW081240220326
41597CB00023BA/4338